I0499720

ARDUINO
AUDIO
PROJECTS

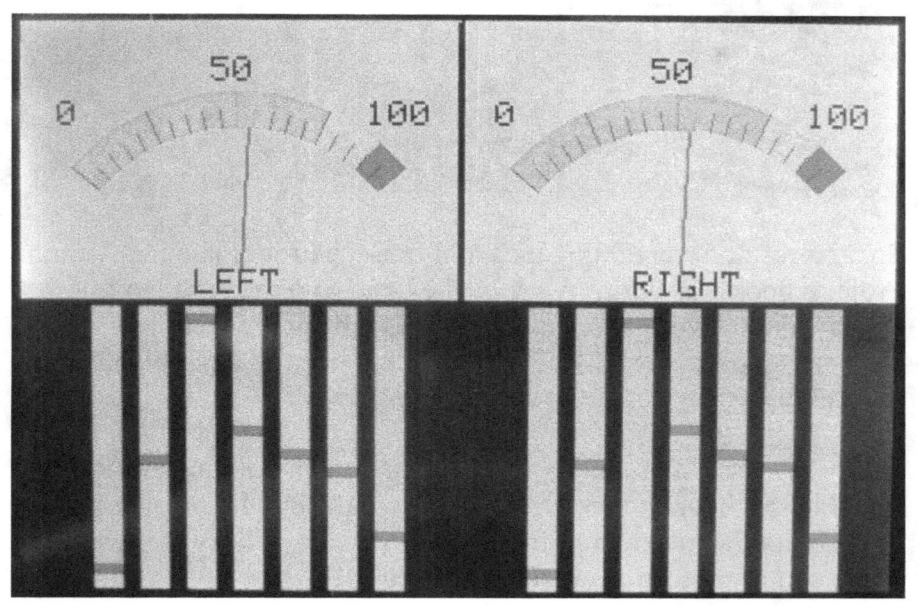

Robert J Davis II

Arduino Audio Projects Copyright 2019 Robert J Davis II

A while back I was rebuilding a Lab Series LS-800 power amplifier with new "guts", a PR-800 power amplifier kit, to be exact. When I was done there were several left over holes in the amplifiers front panel. The amplifier had two 12 step LED VU meters that I had replaced with some LM3915 driver IC's with only 10 LED's each. But what do I do with all of the extra holes? One large rectangular hole was the perfect size for a two-line text type of LCD display. The only way to run such a display would be with an Arduino!

Once that problem was solved, what to do with the six analog inputs of the Arduino was the next issue. You could monitor the audio level, the temperature of the output transistors, or if you added a MSGEQ7, you could display a seven-frequency spectrum analyzer.

This is a picture of the front of the rebuilt audio amplifier.

So, as a result of rebuilding the amplifier, the idea of another Arduino Projects book was born. What can be done with the Arduino to display audio levels, and with a MSGEQ7, display spectrum analysis? You can use LED's, LCD's, and even addressable LED's to display a variety of audio information.

As always, the safe building and operating of any of the devices found in this book are the responsibility of the builder. Connecting the output of a power amplifier into an Arduino will fry the Arduino. Some sort of protection circuit should be used.

Feel free to enhance, rearrange, and otherwise improve on these designs! They are only meant to be a starting point to be built upon and a basis for more complicated devices.

Table of Contents:

Chapter 1

MSGEQ7 Seven Band

Audio Analyzer

Several projects in this book will use the MSGEQ7 Audio Analyzer IC. The MSGEQ7 gives you the ability to display more than just the volume. You can display seven different frequency bands of the audio spectrum. You can make a basic audio spectrum analyzer using an Arduino UNO a MSGEQ7 and some type of display. This audio spectrum display can be as big as you want, from a few LED's to some five-foot-long addressable LED strips.

The MSGEQ7 is a little eight pin chip that has a clock generator, seven band pass amplifiers, seven peak detectors, and an output multiplexer. It is amazing that someone thought to stuff all of that technology into one small IC. The next picture shows the MSGEQ7 pin assignments.

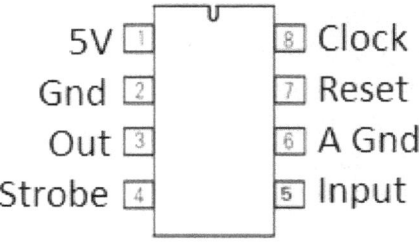

Up next is a picture of the seven frequency bands of the MSGEQ7 taken from the specification sheet. Notice that there is some overlap of the frequencies. The bands are not as narrow as some better spectrum analyzers.

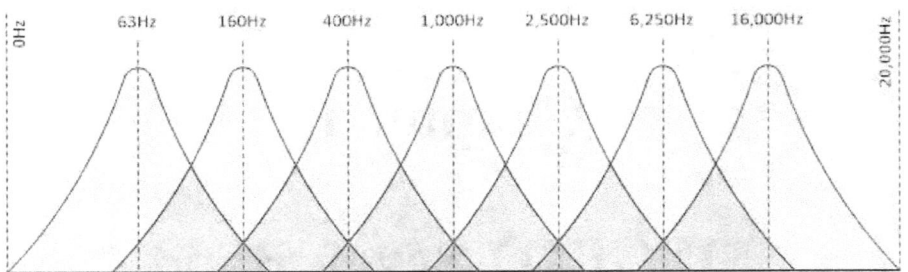

After sending the MSGEQ7 a "Reset" signal the analog output is the 63 Herz frequency band level. Then on each "Strobe" pulse the analog output switches to the next higher frequency band until it reaches the 16KHz band. On the next pulse it rolls over to the lowest frequency. If lack of pins neccesiates it, the Arduino Reset signal could be used and then the programmer has to be careful that the strobe count does not vary from seven or get confused so the wrong band would be selected.

Here is a typical subroutine to read the contents of the MSGEQ7. Note the delay of 30 microseconds for the analog data to stabilize is required after each strobe pulse.

```
void readMSGEQ7() {
  //reset the chip
  digitalWrite(PIN_RESET, HIGH);
  digitalWrite(PIN_RESET, LOW);
  // Loop thru all 7 bands
  for(int band=0; band < 7; band++) {
    // Go to the next band
    digitalWrite(PIN_STROBE,LOW);
    // Delay for data to stabilize
    delayMicroseconds(30);
    // Store left band reading
    left[band] = analogRead(APIN0)/32;
    // Store right band reading
    right[band] = analogRead(APIN1)/32;
    // Reset the strobe pin
    digitalWrite(PIN_STROBE,HIGH);
  }
}
```

There used to be lots of MSGEQ7 shields available on eBay, but they seem to have almost disappeared. I modified the MSGEQ7 shield that I purchased. I changed the input resistors (R3 and R4) from 22K to 1K or 100 ohms. Some designs will even leave those resistors out altogether replacing them with a short.

The lower input resistance works better with headphone level inputs and helps reduce the amount of noise picked up due to the lower impedance. The MSGEQ7 has a built-in protection circuit to protect it from audio levels that are too loud. This next picture shows a typical MSGEQ7 shield. Some newer versions of this shield have a prototype area.

MSGEQ7 Schematic

Up next is the MSGEQ7 Shield schematic diagram. This is the schematic of the Chinese shield; it has a different pin arrangement from the American version. Several projects in this book will need other pin arrangements. Remember to change the 22K resistors to something smaller!

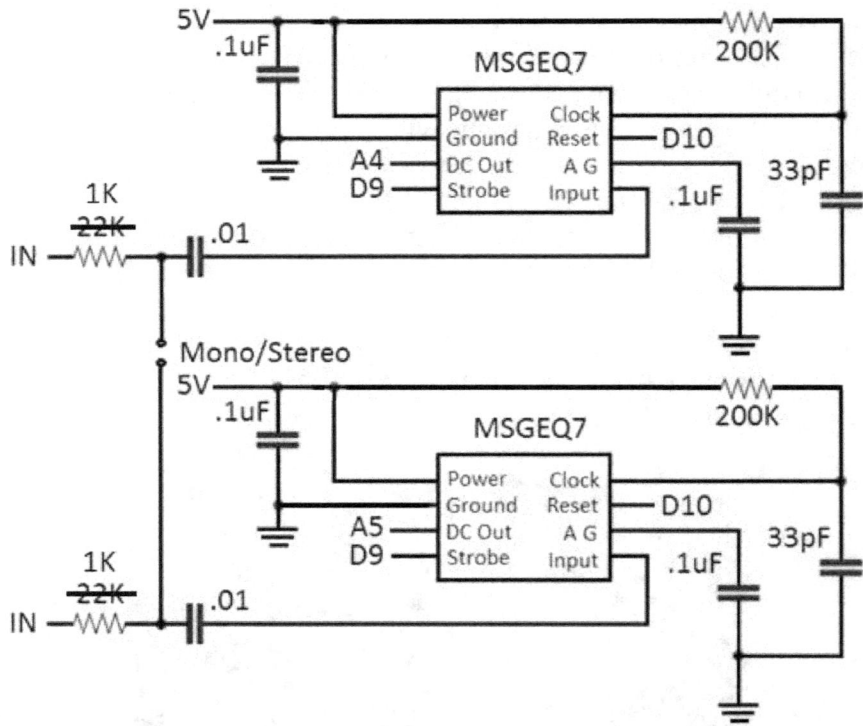

Another option is to buy a MSGEQ7 module like the one that is shown in the next picture. This module does not have the 22K resistors on the input making it much more sensitive. The audio input goes directly to the MSGEQ7 inputs, after going through the input capacitor, relying solely on the internal overvoltage protection.

The MSGEQ7 module does require a six pin connector and cable to use it, or you can just use six jumper wires. The pins are clearly labeled to make hooking it up easier.

The idea of using a module instead of a shield works best for most of the projects found in this book because what pins are available for the MSGEQ7 varies from one project to another project. Changing the pins on the shield is not an easy thing to do. Changing the pins assigned to the MSGEQ7(s) is easier with a module, you just connect the wires to match up with the software.

The biggest complaint about the MSGEQ7 is that it only has seven frequency bands. Most good audio analyzers have as many as 10, 15, or more frequency bands. In the MSGEQ7 each band is the previous bands frequency multiplied by roughly 2.5. However, it is missing the 25 hertz band at the lower end. We could always add that band with a second MSGEQ7 IC that is shifted in frequency.

By changing the 200K resistor, that is used to set the clock frequency in the schematic, to 120 or 150K for higher frequencies, or change the 200K resistor to 270K for lower frequencies, you can offset the detected frequency bands. You can even use three MSGEQ7's with two that are offset in frequency, one higher and one lower, to get a total of 21 frequency bands. This arrangement is shown in the next schematic.

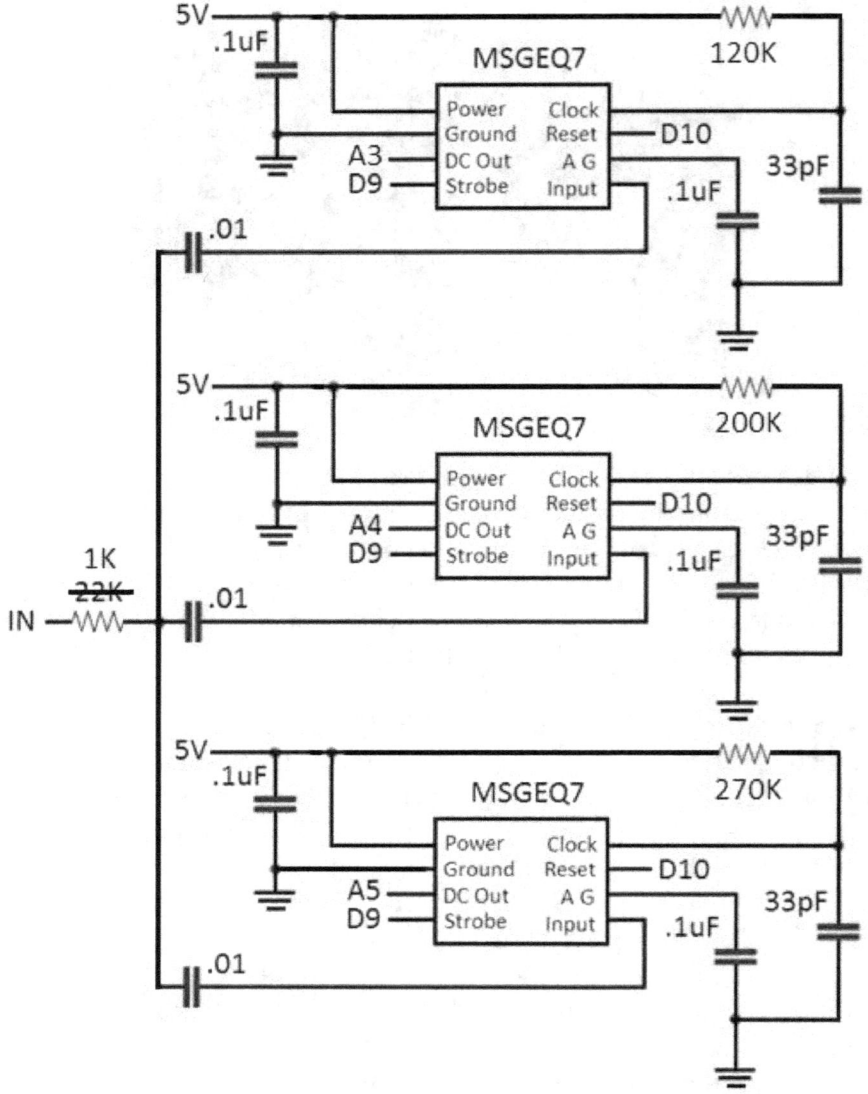

Another way to get more bands is to save the previous samples. You save the current samples and then get another set of samples so now you have 14 bands but they are just delayed in time. This can be done multiple times to get even more samples.

If you want to get even fancier you can search the samples for the peak output of each frequency and light a LED to indicate the most recent peak output level.

Chapter 2

Arduino LED

Audio Projects

Why use an Arduino, when a LM3915 IC can produce a 10 step LED based VU meter with no software? For one thing the Arduino can have more steps, like 14 steps, and it can have up to six inputs and outputs, but not both at the same time. Some of the Arduinos output pins are needed to be used to select the LED column.

Driver transistors like the 2N3904 or a ULN2003 driver IC can be used to select what column is lit up at any one point in time. Besides running individual LED's, this design can also be used with a 5x7 or 8x8 LED array to make the LED spacing more consistent.

You can use 100 ohm or 220 ohm current limiter resistors for the LED's. If you were to attach one of those to D0 you will not be able to communicate with the Arduino. So D0 and D1 were used as the column selection drivers because the 1K resistor in series limits the current to where it does not interfere with communications.

Analog inputs for this design are on analog A0 and A1 inputs. Be sure to limit the input current if the signal strength is greater than five volts. Some power amplifiers output 100 volts! Use a 100 ohm or 1K ohm resistor in series and a five-volt Zener to ground to limit the input voltage and protect the Arduino.

For a fancier interface the audio could go through a coupling capacitor then a 100K resistor to ground. Either way adding a series resistor and a Zener diode to ground will protect the analog input from almost any spikes in power. This next schematic diagram shows how to make a simple front end with overvoltage protection.

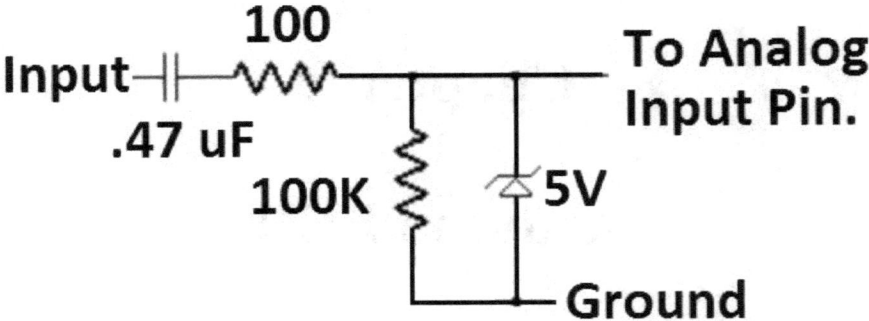

Stereo VU Meter

This is the Arduino and LED display schematic diagram for use with this project.

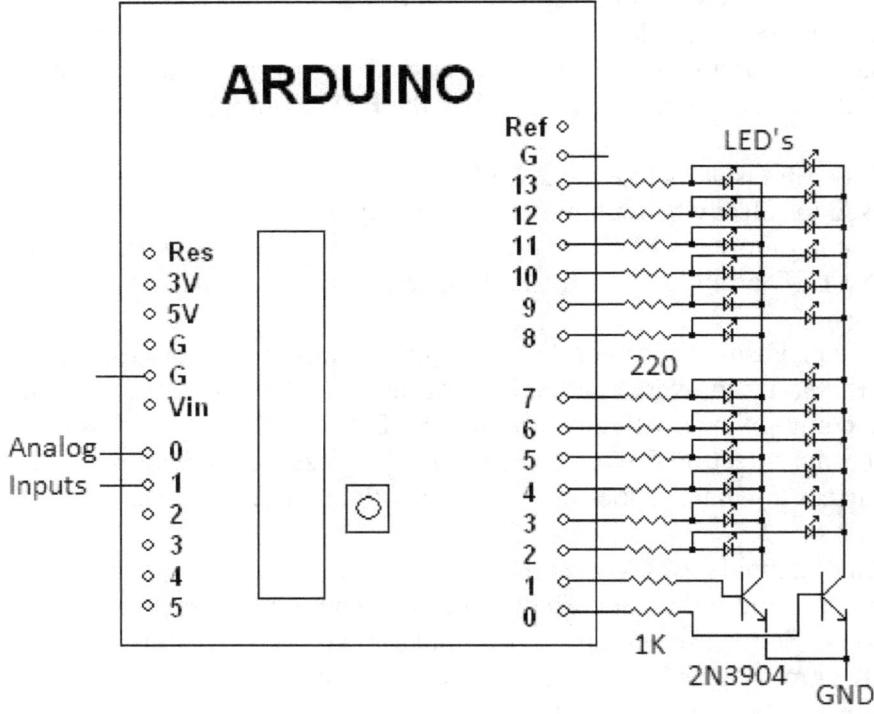

To test out the design, I used a larger sized breadboard. The ground buss on the opposite sides of the breadboard corresponds with the two columns made up of the negative sides of the LED's. Initially I provided a ground to each column until all the LED's were working, then I added the 2N3904 transistors to select between the two columns.

The breadboard test circuit can be seen in this picture.

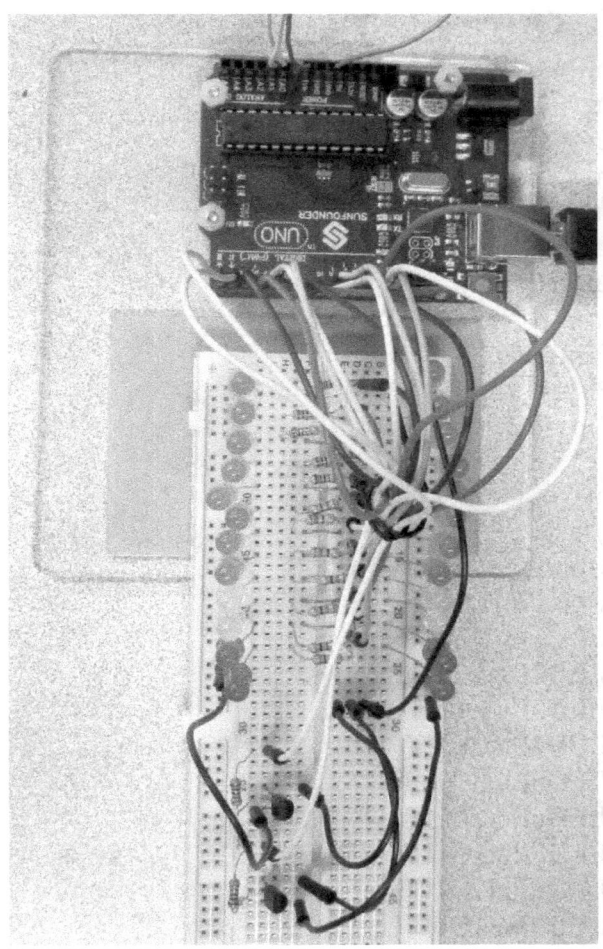

Here is the code to make the two column 24 LED display work. You could alternately create the Data pins as an "array" and then loop through them. That would reduce this code to about half of the current size. I coded this way to make what the software is doing easier for readers to understand.

```
//************************************************/
// Name    : Dual LED VU Meter
// Author  : Bob Davis
// Date    : 8-16-2019
```

```
//  Version : 1.0
//***********************************************/
// Pins for the LED's
int Col0Pin = 0;
int Col1Pin = 1;
int LED1Pin = 2;
int LED2Pin = 3;
int LED3Pin = 4;
int LED4Pin = 5;
int LED5Pin = 6;
int LED6Pin = 7;
int LED7Pin = 8;
int LED8Pin = 9;
int LED9Pin = 10;
int LED10Pin = 11;
int LED11Pin = 12;
int LED12Pin = 13;

// Set the pins to output to the LED's
void setup() {
  pinMode(Col0Pin, OUTPUT);
  pinMode(Col1Pin, OUTPUT);
  pinMode(LED1Pin, OUTPUT);
  pinMode(LED2Pin, OUTPUT);
  pinMode(LED3Pin, OUTPUT);
  pinMode(LED4Pin, OUTPUT);
  pinMode(LED5Pin, OUTPUT);
  pinMode(LED6Pin, OUTPUT);
  pinMode(LED7Pin, OUTPUT);
  pinMode(LED8Pin, OUTPUT);
  pinMode(LED9Pin, OUTPUT);
  pinMode(LED10Pin, OUTPUT);
  pinMode(LED11Pin, OUTPUT);
  pinMode(LED12Pin, OUTPUT);
}

void loop() {
  for (int chan = 0; chan < 2; chan++){
    int AV0=analogRead(A0)/32;
    int AV1=analogRead(A1)/32;
    digitalWrite (LED1Pin, LOW);
    digitalWrite (LED2Pin, LOW);
```

```
digitalWrite (LED3Pin, LOW);
digitalWrite (LED4Pin, LOW);
digitalWrite (LED5Pin, LOW);
digitalWrite (LED6Pin, LOW);
digitalWrite (LED7Pin, LOW);
digitalWrite (LED8Pin, LOW);
digitalWrite (LED9Pin, LOW);
digitalWrite (LED10Pin, LOW);
digitalWrite (LED11Pin, LOW);
digitalWrite (LED12Pin, LOW);

if (chan==0) {
  digitalWrite (Col0Pin, HIGH);
  digitalWrite (Col1Pin, LOW);
  if (AV0 > 1) digitalWrite (LED1Pin, HIGH);
  if (AV0 > 2) digitalWrite (LED2Pin, HIGH);
  if (AV0 > 3) digitalWrite (LED3Pin, HIGH);
  if (AV0 > 4) digitalWrite (LED4Pin, HIGH);
  if (AV0 > 5) digitalWrite (LED5Pin, HIGH);
  if (AV0 > 6) digitalWrite (LED6Pin, HIGH);
  if (AV0 > 7) digitalWrite (LED7Pin, HIGH);
  if (AV0 > 8) digitalWrite (LED8Pin, HIGH);
  if (AV0 > 9) digitalWrite (LED9Pin, HIGH);
  if (AV0 > 10) digitalWrite (LED10Pin, HIGH);
  if (AV0 > 11) digitalWrite (LED11Pin, HIGH);
  if (AV0 > 12) digitalWrite (LED12Pin, HIGH);
}

if (chan==1) {
  digitalWrite (Col1Pin, HIGH);
  digitalWrite (Col0Pin, LOW);
  if (AV1 > 1) digitalWrite (LED1Pin, HIGH);
  if (AV1 > 2) digitalWrite (LED2Pin, HIGH);
  if (AV1 > 3) digitalWrite (LED3Pin, HIGH);
  if (AV1 > 4) digitalWrite (LED4Pin, HIGH);
  if (AV1 > 5) digitalWrite (LED5Pin, HIGH);
  if (AV1 > 6) digitalWrite (LED6Pin, HIGH);
  if (AV1 > 7) digitalWrite (LED7Pin, HIGH);
  if (AV1 > 8) digitalWrite (LED8Pin, HIGH);
  if (AV1 > 9) digitalWrite (LED9Pin, HIGH);
  if (AV1 > 10) digitalWrite (LED10Pin, HIGH);
  if (AV1 > 11) digitalWrite (LED11Pin, HIGH);
```

```
      if (AV1 > 12) digitalWrite (LED12Pin, HIGH);
    }
  delay(5);
  }
}
```

MSGEQ7 and LED Array

You can display even more information with a LED array. An array
is like the two columns in the previous project, but there are five or
even eight columns and they are all contained together in the same
plastic assembly. There are many 8x8 LED arrays around but you
can also make an 8x8 array with two 5x8 LED arrays chained
together. These arrays are available in single color versions, dual
color versions and three-color versions as well.

This next picture compares an 8x8 LED array to some ordinary
LED's. As you can see, the array is much more compact in size and
is much easier to work with than trying to wire up 64 individual
LED's.

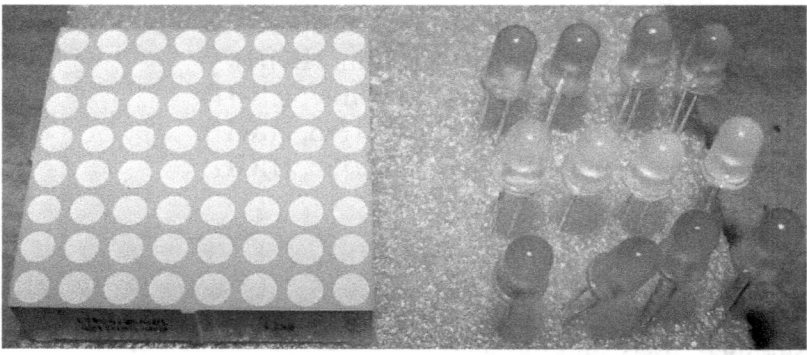

This next image is the schematic diagram of how to connect the array
to and Arduino UNO. I did not give the pinout of the array because
the pinout varies drastically from one array to another. The pin
arrangements vary even more when you try to use two and three color
LED arrays. You can figure out the pin arrangement of an LED array
with a nine-volt battery and a 1K resistor. Basically, you use trial
and error to light up the LED's and record what pin combinations
light what LED's.

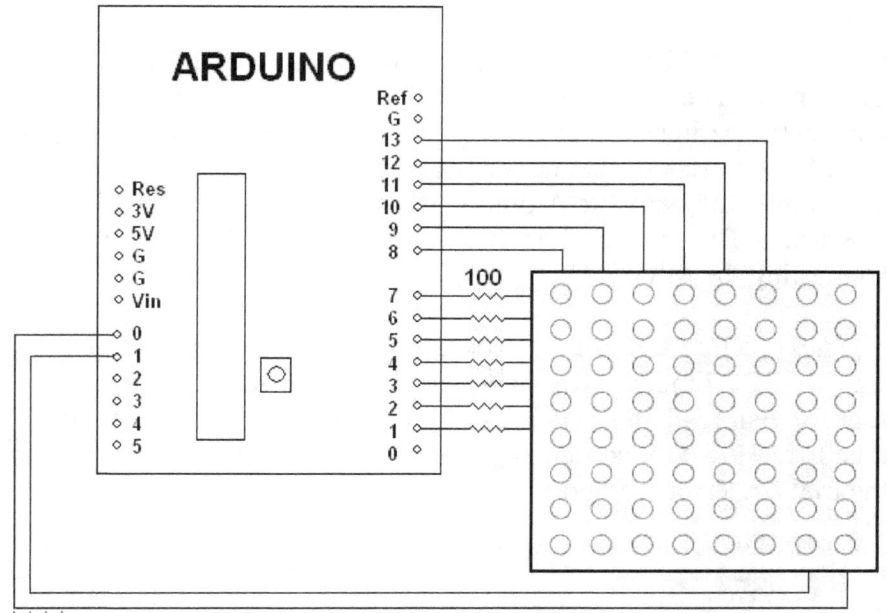

For adding the MSGEQ7 connect strobe to A2, reset to A3 and the analog inputs to A4 and A5 as in shown this schematic.

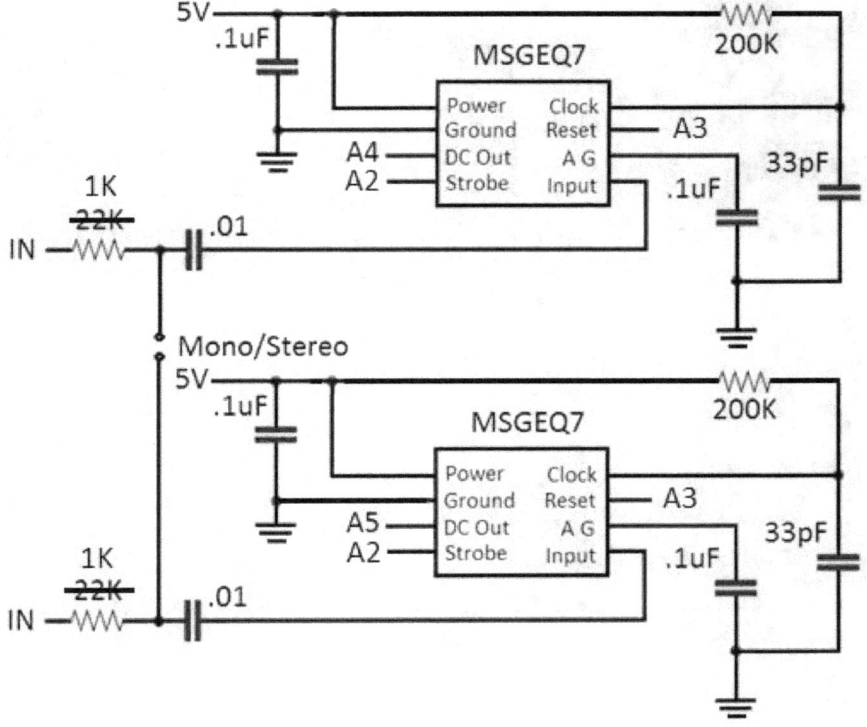

I used two 5x8 LED arrays for my setup. That gives me up to a 10x8 LED array. The top two rows are not used as there are not enough pins on the Arduino UNO to use them, unless you add something like a 74LS138 to reduce the number of Arduino pins that are needed. The number of available Arduino UNO pins limits me to a using a 7x8 array. The other limit is that we need to keep four pins free for connecting the MSGEQ7.

The next picture shows my test setup for this LED Array project. The MSGEQ7 module was hanging off the left side in the picture so it is not visible. I had to use several jumpers to link the LED arrays together. How to link them together was not easy to figure out using the trial and error method.

To troubleshoot this setup I changed the delay to 500 so I could see each column light up one column at a time. Then I connected the analog inputs to five volts so they were maxed out. You could also use variable resistors. That way you can work with the LED array untill all of the LED's are lit in the correct order.

This is the code to run an 7x8 LED array with a MSGEQ7.

```
//*****************************************//
// Name   : Breadboard 7x10 MSGEQ7      //
// Author : Bob Davis                //
// Date   : 10 Oct, 2019             //
//*****************************************//

// Pins for the row drivers
int colPin1 = 1;
int colPin2 = 2;
int colPin3 = 3;
int colPin4 = 4;
int colPin5 = 5;
int colPin6 = 6;
int colPin7 = 7;

// Without 74LS138
int rdataPin1 = 8;
int rdataPin2 = 9;
int rdataPin3 = 10;
int rdataPin4 = 11;
int rdataPin5 = 12;
int rdataPin6 = 13;
int rdataPin7 = 14;
int rdataPin8 = 15;

// store MSGEQ7 values here
int left[7];
int right[7];
#define PIN_STROBE 16 // Analog 2
#define PIN_RESET 17  // Analog 3
#define APIN0 4 //analog input
#define APIN1 5 //analog input

void readMSGEQ7() { //reset the chip
```

```
    digitalWrite(PIN_RESET, HIGH);
    digitalWrite(PIN_RESET, LOW);
    for(int band=0; band < 7; band++) {  // loop thru all 7 bands
      digitalWrite(PIN_STROBE,LOW);      // go to the next band
      delayMicroseconds(30);             // gather data
      left[band] = analogRead(APIN0)/32;   // store band reading
      right[band] = analogRead(APIN1)/32;  // store band reading
      digitalWrite(PIN_STROBE,HIGH);     // reset the strobe pin
    }
}

// Set the pins to output to the sign
void setup() {
  pinMode(PIN_STROBE, OUTPUT);
  pinMode(PIN_RESET, OUTPUT);

  pinMode(colPin1, OUTPUT);
  pinMode(colPin2, OUTPUT);
  pinMode(colPin3, OUTPUT);
  pinMode(colPin4, OUTPUT);
  pinMode(colPin5, OUTPUT);
  pinMode(colPin6, OUTPUT);
  pinMode(colPin7, OUTPUT);

  pinMode(rdataPin1, OUTPUT);
  pinMode(rdataPin2, OUTPUT);
  pinMode(rdataPin3, OUTPUT);
  pinMode(rdataPin4, OUTPUT);
  pinMode(rdataPin5, OUTPUT);
  pinMode(rdataPin6, OUTPUT);
  pinMode(rdataPin7, OUTPUT);
  pinMode(rdataPin8, OUTPUT);
}

void loop(){
  // select the row
  readMSGEQ7();
  for (int row = 0; row < 7; row++)  {
    // Turn Display off
    digitalWrite (colPin1, LOW);
    digitalWrite (colPin2, LOW);
    digitalWrite (colPin3, LOW);
```

```
digitalWrite (colPin4, LOW);
digitalWrite (colPin5, LOW);
digitalWrite (colPin6, LOW);
digitalWrite (colPin7, LOW);

digitalWrite(rdataPin1, HIGH);
digitalWrite(rdataPin2, HIGH);
digitalWrite(rdataPin3, HIGH);
digitalWrite(rdataPin4, HIGH);
digitalWrite(rdataPin5, HIGH);
digitalWrite(rdataPin6, HIGH);
digitalWrite(rdataPin7, HIGH);
digitalWrite(rdataPin8, HIGH);

// Display the data results LOW = "on"
if (right[row]>1 ) digitalWrite(rdataPin1, LOW);
if (right[row]>2 ) digitalWrite(rdataPin2, LOW);
if (right[row]>3 ) digitalWrite(rdataPin3, LOW);
if (right[row]>4 ) digitalWrite(rdataPin4, LOW);
if (right[row]>5 ) digitalWrite(rdataPin5, LOW);
if (right[row]>6 ) digitalWrite(rdataPin6, LOW);
if (right[row]>7 ) digitalWrite(rdataPin7, LOW);
if (right[row]>8 ) digitalWrite(rdataPin8, LOW);

// Select row to turn on HIGH = "on"
if (row==0) digitalWrite (colPin1, HIGH);
if (row==1) digitalWrite (colPin2, HIGH);
if (row==2) digitalWrite (colPin3, HIGH);
if (row==3) digitalWrite (colPin4, HIGH);
if (row==4) digitalWrite (colPin5, HIGH);
if (row==5) digitalWrite (colPin6, HIGH);
if (row==6) digitalWrite (colPin7, HIGH);

// Wait to see what we sent to the display ;
delay(1);
 }
}
```

Chapter 3

Arduino Text LCD

Audio Projects

Our next project will be to make a two-line 1602 text LCD display operational. Then we will make it display some simple analog or audio bar graphs by creating custom characters. Initially we will have two analog inputs for right and left channel. Then we will add a MSGEQ7 input and some new software making the 1602 LCD into a 14-frequency bar graph.

The Arduino analog input voltage has 10 bits for a maximum value of 1024. We divide that input voltage by 2.05 to get a maximum of 5000 millivolts or 5 Volts. Then we divide that result by three to get a maximum of 166.6. Next we divide that number by 10 to get a maximum of 16.66. There are only 16 available character positions, so the last .66 is lost as off the right side of our bar graph scale. Thus, the display will max out at about 4.8 volts.

Besides the bar graphs the LCD will show the input voltage in millivolts. This is overwritten when the bar graph reaches the right side of the LED display.

On the next page there is a picture of the stereo analog or audio LCD display showing it running. The design and testing of this project were done with two variable resistors so the response would be slower and it would be easier to monitor the display and detect any errors.

Not visible in the picture is the ground wire to the LCD RW pin five and the wires to and from the contrast trimming variable resistor. Those wires are all hard soldered on the back side of the LCD display to make it easier to wire it up.

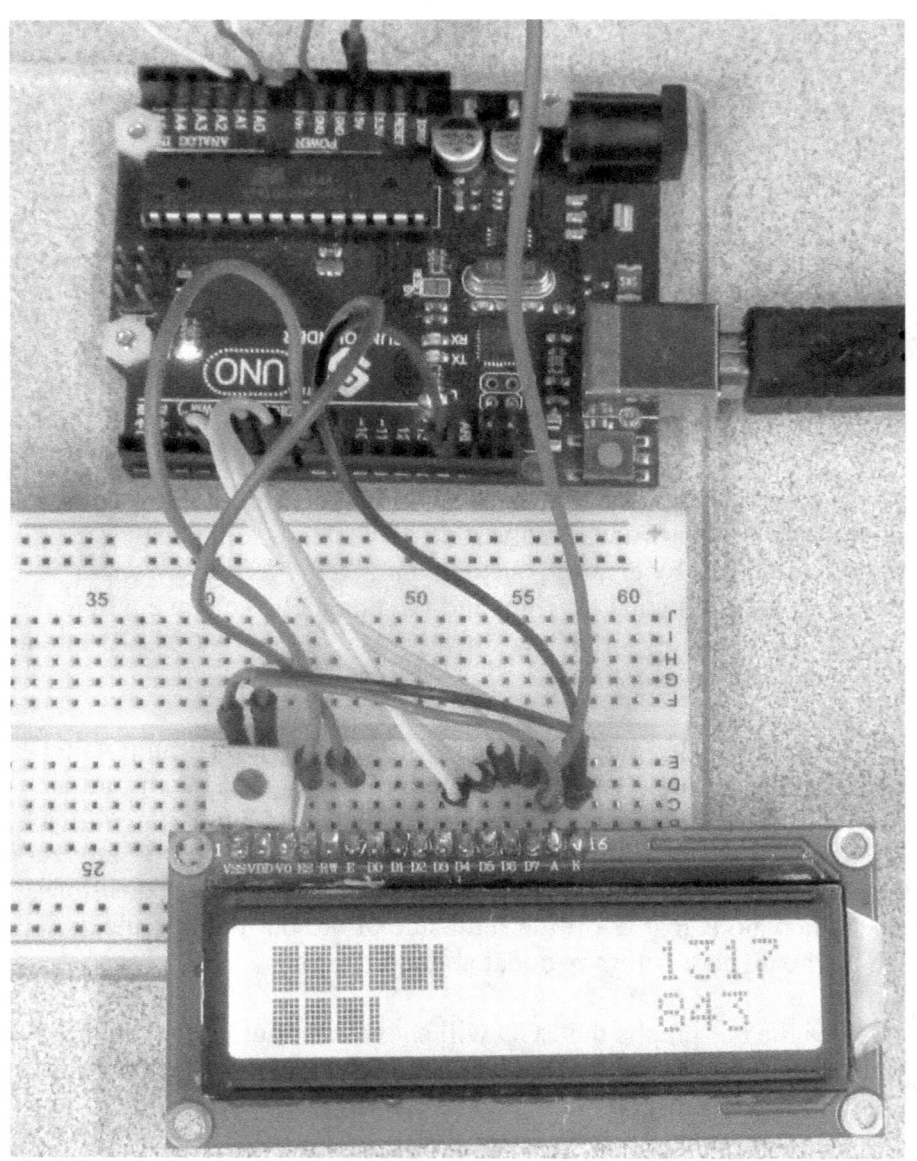

Below is the schematic diagram of the two-line LCD interface. It is different from previous versions of 1602 LCD schematics found in some of my earlier books. The changes were made in order to be compatible with the 1602 LCD shield that is now very commonly found on eBay.

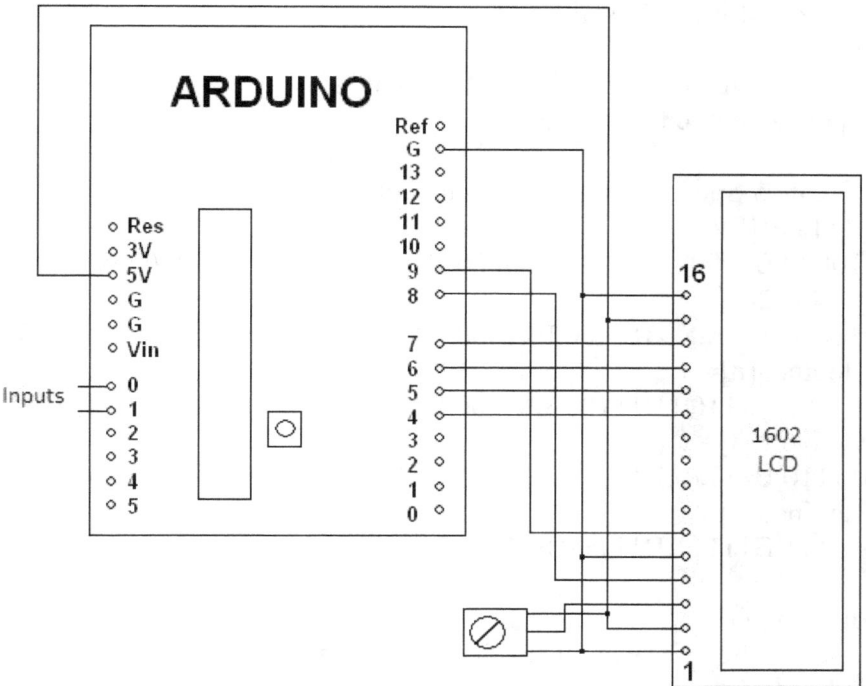

Here is the sketch or code for a stereo audio meter display. It uses custom characters to create up to five lines for the five bar widths inside of each character position. I used variable resistors for the input to test this sketch. The variable resistors supplied input voltages that varied from zero volts to five volts.

```
//*****************************************
// 1602 Audio Meter
// 8-19-2019 By Bob Davis
// Wiring:
//  LCD RS pin to digital pin 8
//  LCD EN pin to digital pin 9
//  LCD D4 pin to digital pin 4
//  LCD D5 pin to digital pin 5
//  LCD D6 pin to digital pin 6
```

```
// LCD D7 pin to digital pin 7
// LCD R/W pin to ground
// Variable resistor wiper to LCD VO pin (pin 3)
//

// include the library code:
#include <LiquidCrystal.h>

// initialize the library with interface pin #'s
LiquidCrystal lcd(8, 9, 4, 5, 6, 7);

// Create 5 Special characters being a line from left to right:
byte line1[8]= {
B10000,B10000,B10000,B10000,B10000,B10000,B10000,B10000};
byte line2[8]= {
B11000,B11000,B11000,B11000,B11000,B11000,B11000,B11000};
byte line3[8]= {
B11100,B11100,B11100,B11100,B11100,B11100,B11100,B11100};
byte line4[8]= {
B11110,B11110,B11110,B11110,B11110,B11110,B11110,B11110};
byte line5[8]= {
B11111,B11111,B11111,B11111,B11111,B11111,B11111,B11111};

void setup() {
  // set up the LCD's number of columns and rows:
  lcd.begin(16, 2);
  // Send the new characters to the LCD.
  lcd.createChar(1, line1);
  lcd.createChar(2, line2);
  lcd.createChar(3, line3);
  lcd.createChar(4, line4);
  lcd.createChar(5, line5);
}

void loop() {
  lcd.clear();
  int Voltage0=analogRead(A0)/2.05;
  int Voltage1=analogRead(A1)/2.05;
  // Optionally print the voltage in mv:
  lcd.setCursor(12, 0);
  lcd.print(Voltage0);
  lcd.setCursor(12, 1);
```

```
lcd.print(Voltage1);

int metervolts=Voltage0/3;
for(int pos=0; pos<metervolts/10; pos++){
  lcd.setCursor(pos, 0);
  lcd.write(255);
  }
lcd.setCursor(metervolts/10, 0);
if (metervolts%10/2 == 0) lcd.write(1);
if (metervolts%10/2 == 1) lcd.write(2);
if (metervolts%10/2 == 2) lcd.write(3);
if (metervolts%10/2 == 3) lcd.write(4);
if (metervolts%10/2 == 4) lcd.write(5);
metervolts=Voltage1/3;
for(int pos=0; pos<metervolts/10; pos++){
  lcd.setCursor(pos, 1);
  lcd.write(255);
  }
lcd.setCursor(metervolts/10, 1);
if (metervolts%10/2 == 0) lcd.write(1);
if (metervolts%10/2 == 1) lcd.write(2);
if (metervolts%10/2 == 2) lcd.write(3);
if (metervolts%10/2 == 3) lcd.write(4);
if (metervolts%10/2 == 4) lcd.write(5);
delay(200);
}
```

MSGEQ7 Bar Graph

For the next project we will add two MSGEQ7's and turn the bar graphs sideways so we can easily display 14 bar graphs to correspond with the 14 frequency bands that are output from the two MSGEQ7's.

Besides being turned sideways, the bar graphs will be split in half so that an analog input value of 0-7 will appear on the bottom row of what is normally characters and an input value of 8-15 will appear on the top row of what is normally characters of the 1602 LCD.

The 1602 schematic is the same as before, but the MSGEQ7 signals are added in four places. The MSGEQ7 Strobe in on pin 10 and the Reset is on pin 11. Analog inputs 0 and 1 are used for the analog values coming from the MSGEQ7. This can be seen in the next schematic.

This is the MSGEQ7 schematic for this project.

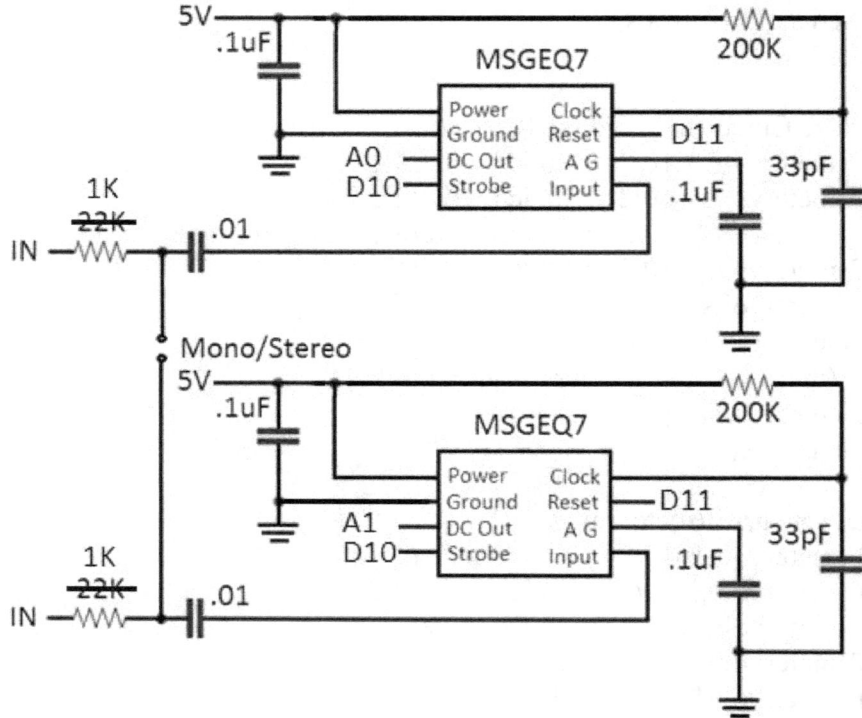

Once again, I tested the setup with variable resistors connected to analog inputs 0 and 1 to get the display working properly before connecting up the MSGEQ7.

This next picture is of my arrangement for using the 1602 LCD display to show the outputs of the two MSGEQ7's. The MSGEQ7 module is seen from the end view as it is plugged directly into the breadboard. As a result, the MSGEQ7 module is not very visible in the picture.

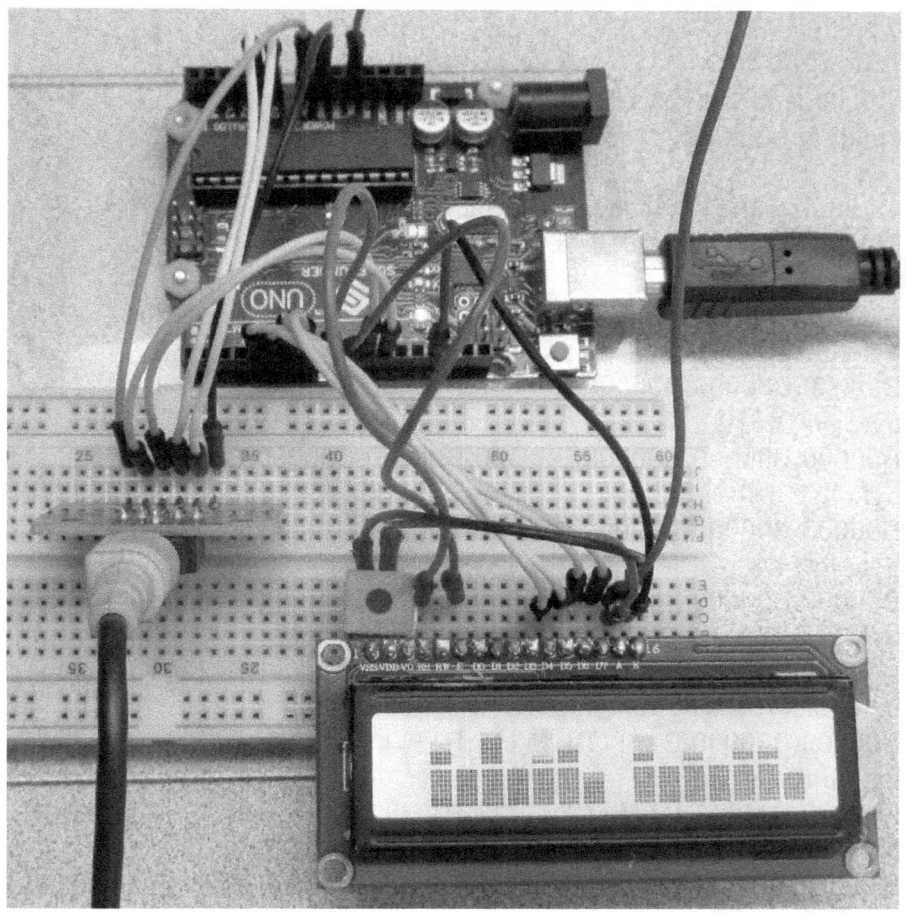

Here is the code to get the 1602 to work with the MSGEQ7. It also programs custom characters that are made of up to eight horizontal lines.

```
//*****************************************
// 1602 MSGEQ7 Audio Meter
// 8-19-2019 By Bob Davis
// Wiring:
// * LCD RS pin to digital pin 8
// * LCD En pin to digital pin 9
// * LCD D4 pin to digital pin 4
// * LCD D5 pin to digital pin 5
// * LCD D6 pin to digital pin 6
// * LCD D7 pin to digital pin 7
// * LCD R/W pin to ground
// * Variable resistor wiper to LCD VO pin (pin 3)
```

```
//

// include the library code:
#include <LiquidCrystal.h>

// initialize the library with interface pin #'s
LiquidCrystal lcd(8, 9, 4, 5, 6, 7);

// Create 8 Special characters being a line from bottom to top:
byte line1[8] = {
B00000,B00000,B00000,B00000,B00000,B00000,B00000,B11111};
byte line2[8] = {
B00000,B00000,B00000,B00000,B00000,B00000,B11111,B11111};
byte line3[8] = {
B00000,B00000,B00000,B00000,B00000,B11111,B11111,B11111};
byte line4[8] = {
B00000,B00000,B00000,B00000,B11111,B11111,B11111,B11111};
byte line5[8] = {
B00000,B00000,B00000,B11111,B11111,B11111,B11111,B11111};
byte line6[8] = {
B00000,B00000,B11111,B11111,B11111,B11111,B11111,B11111};
byte line7[8] = {
B00000,B11111,B11111,B11111,B11111,B11111,B11111,B11111};
byte line8[8] = {
B11111,B11111,B11111,B11111,B11111,B11111,B11111,B11111};

// store MSGEQ7 values here
int left[7];
int right[7];
#define PIN_STROBE 10
#define PIN_RESET 11
#define APIN0 0 //analog input
#define APIN1 1 //analog input

void readMSGEQ7() { //reset the chip
  digitalWrite(PIN_RESET, HIGH);
  digitalWrite(PIN_RESET, LOW);
  for(int band=0; band < 7; band++) {  // loop thru all 7 bands
    digitalWrite(PIN_STROBE,LOW);      // go to the next band
    delayMicroseconds(30);             // gather data
    left[band] = analogRead(APIN0)/32;   // store band reading
    right[band] = analogRead(APIN1)/32;  // store band reading
```

```
    digitalWrite(PIN_STROBE,HIGH);     // reset the strobe pin
    }
}

void setup() {
  // set up the LCD's number of columns and rows:
  lcd.begin(16, 2);
  // Send the new characters to the LCD.
  lcd.createChar(1, line1);
  lcd.createChar(2, line2);
  lcd.createChar(3, line3);
  lcd.createChar(4, line4);
  lcd.createChar(5, line5);
  lcd.createChar(6, line6);
  lcd.createChar(7, line7);
  lcd.createChar(8, line8);
  pinMode(PIN_STROBE, OUTPUT);
  pinMode(PIN_RESET, OUTPUT);
}

void loop() {
  lcd.clear();
  readMSGEQ7();
//  lcd.write("L      R");
  lcd.setCursor(1, 0);
  for(int pos=0; pos<7; pos++){
    if (left[pos]>8) lcd.write(left[pos]-8);
    }
  lcd.setCursor(9, 0);
  for(int pos=0; pos<7; pos++){
    if (right[pos]>8) lcd.write(right[pos]-8);
    }
  lcd.setCursor(1, 1);
  for(int pos=0; pos<7; pos++){
    if (left[pos]<8){
      lcd.write(left[pos]+1);
    }
    else lcd.write(255);
  }
  lcd.setCursor(9, 1);
  for(int pos=0; pos<7; pos++){
    if (right[pos]<8){
```

```
        lcd.write(right[pos]+1);
     }
   else lcd.write(255);
  }
 delay(300);
}
```

Chapter 4

Arduino Small Graphic

LCD Projects

Our next project will be to set up our first graphics type of LCD display. Graphics LCD's are very diverse. Unlike text based LCD's, they do not use the same controller, or the same interface chip. Even the pin connection will vary widely. On the Nokia LCD displays look for a pin with a square pad around it. The square pan usually designates pin number one.

The drivers for graphics LCD's are not included in the Arduino library. You have to download the compressed drivers for each graphics LCD and then extract or unzip the drivers into the "Arduino\libraries" directory. For the Nokia LCD, the driver file is called "Adafruit-PCD8544-Nokia-5110-LCD-library-master".

The next picture is a screen capture is what the Arduino\Libraries directory looks like with the Nokia LCD drivers unzipped into it.

Once all of the drivers are correctly installed, then you can plug in the Arduino interfaced with the Nokia LCD, and hopefully get it to work correctly. There is a example program that should also show up for you to try called "pcdtest".

Six Analog Bar Graph

Up next is the schematic diagram of how to connect the Nokia LCD to the Arduino UNO. Level shifters are not needed even though the logic level for the Nokia display is 3.3 volts, its inputs can tolerate five volts. Even the LED 100 ohm resistor is left out in some designs but I prefer to keep it. Note that the wires to D3 and D4 are swapped. That is done to match up with the Adafruit driver code.

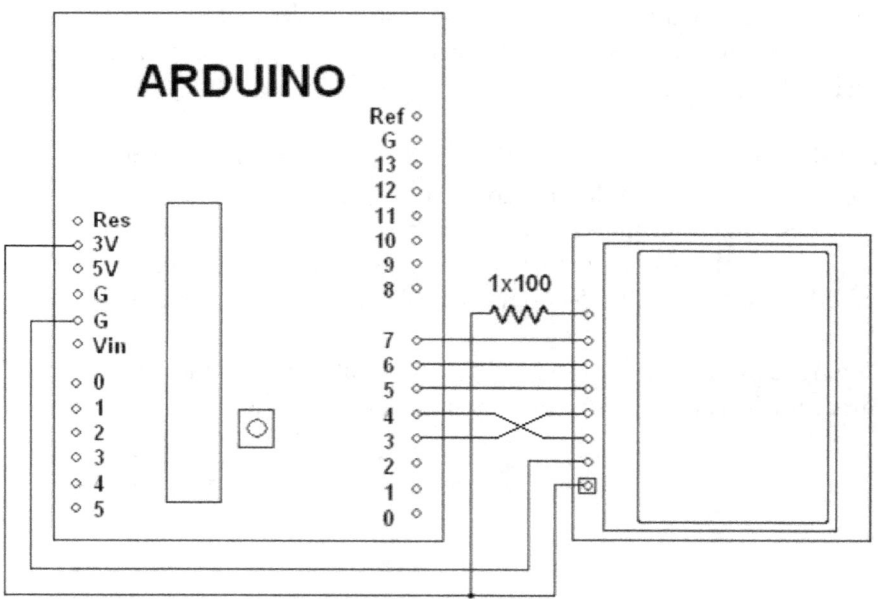

Up next is a picture of the Nokia display working. It displays six values and six bar graphs to match the Arduinos six analog inputs.

This is the code to display the six analog values.

```
/*********************************************
Title:   Nokia 6 Analog Display
By:      Bob Davis
Note:    This code uses the Adafruit PDC8544 LCD library
*********************************************/
#include <SPI.h>
#include <Adafruit_GFX.h>
#include <Adafruit_PCD8544.h>
Adafruit_PCD8544 display = Adafruit_PCD8544(7, 6, 5, 4, 3);

void setup(){
  display.begin();
  display.setContrast(50);
  display.clearDisplay();
```

```
}

void loop(){
display.clearDisplay();
// Set up for text.
display.setTextSize(1);
display.setTextColor(BLACK);
//display.setCursor(0,0);
int analog0 = analogRead(A0)/2.05;
int analog1 = analogRead(A1)/2.05;
int analog2 = analogRead(A2)/2.05;
int analog3 = analogRead(A3)/2.05;
int analog4 = analogRead(A4)/2.05;
int analog5 = analogRead(A5)/2.05;
// display text values
display.setCursor(64,0);
display.println(analog0);
display.setCursor(64,8);
display.println(analog1);
display.setCursor(64,16);
display.println(analog2);
display.setCursor(64,24);
display.println(analog3);
display.setCursor(64,32);
display.println(analog4);
display.setCursor(64,40);
display.println(analog5);
// display lines
display.drawRect(0, 0, analog0/6, 6, BLACK);
display.drawRect(0, 8, analog1/6, 6, BLACK);
display.drawRect(0, 16, analog2/6, 6, BLACK);
display.drawRect(0, 24, analog3/6, 6, BLACK);
display.drawRect(0, 32, analog4/6, 6, BLACK);
display.drawRect(0, 40, analog5/6, 6, BLACK);
//Now push the buffer to the LCD for display
display.display();
delay(200);
}
```

MSGEQ7 Bar Graph

When we add the MSGEQ7 Strobe is 9 and reset is 10. The analog inputs are on A4 and A5. This matches the schematic given back in the chapter on the MSGEQ7.

This is the code to make 14 bar graphs from the outputs of the MSGEQ7.

```
/***********************************************
Title:  Nokia Dual MSGEQ7 Display
By:     Bob Davis
Note:   This code uses the Adafruit PDC8544 LCD library
***********************************************/
#include <SPI.h>
#include <Adafruit_GFX.h>
#include <Adafruit_PCD8544.h>
Adafruit_PCD8544 display = Adafruit_PCD8544(7, 6, 5, 4, 3);
int left[7];
int right[7];
#define PIN_STROBE 9
#define PIN_RESET 10
#define APIN0 4 //analog input
#define APIN1 5 //analog input

void readMSGEQ7() {
  //reset the chip
  digitalWrite(PIN_RESET, HIGH);
  digitalWrite(PIN_RESET, LOW);
  // Loop thru all 7 bands
  for(int band=0; band < 7; band++) {
    // Go to the next band
    digitalWrite(PIN_STROBE,LOW);
    // Delay for data to stabilize
    delayMicroseconds(30);
    // Store left band reading 10 sets 84 as full scale
    left[band] = analogRead(APIN0)/10;
    // Store right band reading
    right[band] = analogRead(APIN1)/10;
    // Reset the strobe pin
    digitalWrite(PIN_STROBE,HIGH);
  }
```

```
}

void setup(){
 pinMode(PIN_STROBE, OUTPUT);
 pinMode(PIN_RESET, OUTPUT);
 display.begin();
 display.setContrast(50);
 display.clearDisplay();
}

void loop(){
 display.clearDisplay();
 // Set up for text.
 display.setTextSize(1);
 display.setTextColor(BLACK);
 //display.setCursor(0,0);
 readMSGEQ7();
 // display lines
 display.drawRect(0, 0, right[0], 2, BLACK);
 display.drawRect(0, 3, right[1], 2, BLACK);
 display.drawRect(0, 6, right[2], 2, BLACK);
 display.drawRect(0, 9, right[3], 2, BLACK);
 display.drawRect(0, 12, right[4], 2, BLACK);
 display.drawRect(0, 15, right[5], 2, BLACK);
 display.drawRect(0, 18, right[6], 2, BLACK);

 display.drawRect(0, 24, left[0], 2, BLACK);
 display.drawRect(0, 27, left[1], 2, BLACK);
 display.drawRect(0, 30, left[2], 2, BLACK);
 display.drawRect(0, 33, left[3], 2, BLACK);
 display.drawRect(0, 36, left[4], 2, BLACK);
 display.drawRect(0, 39, left[5], 2, BLACK);
 display.drawRect(0, 42, left[6], 2, BLACK);
 //Now push the buffer to the LCD for display
 display.display();
 delay(200);
}
```

Chapter 5

Arduino Graphic LCD

Audio Projects

For the next two projects we will add a graphic type of LCD screen. I chose to use an LCD screen that comes in a plug-in shield to make the wiring as easy as is possible. There are at least three sizes of LCD screen shields that plug directly into an Arduino. The smaller screen is 2.4 inches and the next size is 3.2 inches and the largest size is 3.5 inches diagonal. Each size features a different number of pixels. So, the software was written so it can easily be scaled to work with different sizes of screens.

The screen that measures 2.4 inches offers a resolution of 320 by 240. That is the default size supported by the Adafruit LCD drivers. The screen that measures 3.2 inches diagonal has a resolution of 400 horizontal pixels by 240 vertical pixels. The largest screen is 3.5

inches and has a resolution of 480 by 320 pixels. Using one of these screens results in lots of pixels allowing many different ways to display the data.

The 2.4 and the 3.5 inch LCD screens use the same pinout as seen in this next picture.

The 3.2 inch LCD rearranges some of the pins as compared to the 2.4 or 3.5 inch models.

This chart shows the pin assignments for the three sizes of LCD's:

Pin Name	2.4+3.5 LCD	3.2 LCD
GND	GND	GND
3V3	3.3V	3.3V
CS	A3	A3
RS	A2	A2
WR	A1	A1
RD	A0	A0
RST	A4	RESET
LED	GND	GND
DB0	8	8

DB1	9	9
DB2	2	10
DB3	3	11
DB4	4	4
DB7	5	13
DB6	6	6
DB7	7	7

Note that A4 is stolen for the LCD "Reset" signal on the 2.4 and the 3.5-inch screens. That will have to be routed to the Arduino Reset instead to free up that pin for analog inputs. However, these shields do not have a connection to the Arduino reset signal. Also, what pins are available for the MSGEQ7 varies with the LCD pinouts, so there are two different MSGEQ7 wiring setups in this chapter.

Stereo Analog Looking Meter

We can now draw what looks like analog meters on the LCD screen. This next program draws two meters, one for each audio channel. The code is based on an example that is found in Instructables, but it has been rewritten so that it can scale between the multiple screen sizes and modified so that it shows two meters. The second meter is a copy of the first meter, offset by width divided by two to reposition it to the right half of the screen.

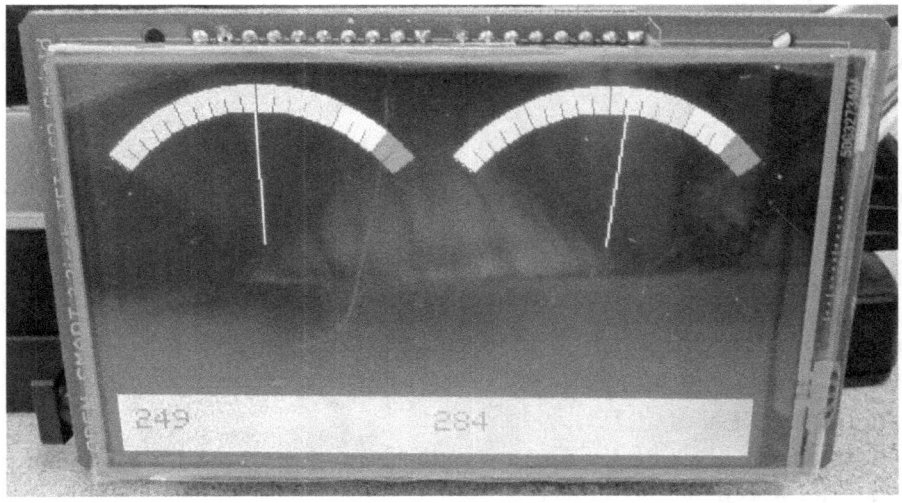

Here is the code for the dual analog meters on the LCD Screen. The code is written to allow for some rescaling to different sizes of

LCD's. It also displays the input voltage in millivolts at the bottom of the screen.

```
// OPEN SMART DUAL METER
// By Bob Davis
// Adafruit_TFTLCD LIBRARY MUST BE CONFIGURED.
// SEE RELEVANT COMMENTS IN Adafruit_TFTLCD.h.
// Some code by Open-Smart Team and Catalex Team
//
// Arduino IDE: 1.8.1
// Board: Arduino UNO R3 or Arduino Mega2560

#include <MCUFRIEND_kbv.h>
MCUFRIEND_kbv tft;
int width=400;  // scales to 320
int height=240;
uint16_t g_identifier;
int lnx0, lny0, lnx1, lny1;
int rnx0, rny0, rnx1, rny1;

//*********************************************//
// Pin assignments for the LCD
// GND          -- GND
// 3V3          -- 3.3V
// CS          -- A3
// RS          -- A2
// WR           -- A1
// RD           -- A0
// RST          -- RESET
// LED          -- GND
// DB0-DB7      -- 8, 9, 10, 11, 4, 13, 6, 7

// Assign names to some common 16-bit color values:
#defineBLACK   0x0000
#defineBLUE    0x001F
#defineRED     0xF800
#defineGREEN   0x07E0
#define CYAN    0x07FF
#define MAGENTA 0xF81F
#define YELLOW  0xFFE0
#define WHITE   0xFFFF
```

```
void setup(void) {
  g_identifier = tft.readID();
  if (g_identifier == 0x00D3 || g_identifier == 0xD3D3) g_identifier =
0x9481;
  // write-only shield
  if (g_identifier == 0xFFFF) g_identifier = 0x9341; // serial
  tft.begin(g_identifier);
  tft.setRotation(1);
  tft.setCursor(0, 0);
  tft.fillRect(0, 0, width, height, BLACK);
}

void loop(void) {
  int rval=analogRead(A4)/2.05;
  int lval=analogRead(A5)/2.05;
  tft.setTextColor(WHITE);
  tft.setTextSize(2);
  tft.setCursor(20,height/3.5);

  //Meter code Based on Sketch From Instructables by Bodmer
  // Draw ticks every 5 degrees from -50 to +50
  for (int i = -50; i < 50; i += 5) {
    // Long scale length
    int tl = 15;
    int w2 = width/2;

    // Start Coodinates of section to draw
    float sx = cos((i - 90) * 0.0174532925);
    float sy = sin((i - 90) * 0.0174532925);
    uint16_t x0 = sx * (100 + tl) + width/4;
    uint16_t y0 = sy * (100 + tl) + height/2;
    uint16_t x1 = sx * 100 + width/4;
    uint16_t y1 = sy * 100 + height/2;

    // Stop Coordinates of section to fill
    float sx2 = cos((i + 5 - 90) * 0.0174532925);
    float sy2 = sin((i + 5 - 90) * 0.0174532925);
    int x2 = sx2 * (100 + tl) + width/4;
    int y2 = sy2 * (100 + tl) + height/2;
    int x3 = sx2 * 100 + width/4;
    int y3 = sy2 * 100 + height/2;
```

```
// Green Section
if (i >= -50 && i < 25) {
  tft.fillTriangle(x0, y0, x1, y1, x2, y2, GREEN);
  tft.fillTriangle(x1, y1, x2, y2, x3, y3, GREEN);
  tft.fillTriangle(x0+w2, y0, x1+w2, y1, x2+w2, y2, GREEN);
  tft.fillTriangle(x1+w2, y1, x2+w2, y2, x3+w2, y3, GREEN);
}

// Yellow Section
if (i >= 25 && i < 40) {
  tft.fillTriangle(x0, y0, x1, y1, x2, y2, YELLOW);
  tft.fillTriangle(x1, y1, x2, y2, x3, y3, YELLOW);
  tft.fillTriangle(x0+w2, y0, x1+w2, y1, x2+w2, y2, YELLOW);
  tft.fillTriangle(x1+w2, y1, x2+w2, y2, x3+w2, y3, YELLOW);
}

// Red Section
if (i >= 40 && i < 55) {
  tft.fillTriangle(x0, y0, x1, y1, x2, y2, RED);
  tft.fillTriangle(x1, y1, x2, y2, x3, y3, RED);
  tft.fillTriangle(x0+w2, y0, x1+w2, y1, x2+w2, y2, RED);
  tft.fillTriangle(x1+w2, y1, x2+w2, y2, x3+w2, y3, RED);
}

// Short left scale tick length
if (i % 25 != 0) tl = 8;
// Calculate coords for short tick length
x0 = sx * (100 + tl) + width/4;
y0 = sy * (100 + tl) + height/2;
x1 = sx * 100 + width/4;
y1 = sy * 100 + height/2;
// Draw tick
tft.drawLine(x0, y0, x1, y1, BLACK);

// Short right scale tick length
if (i % 25 != 0) tl = 8;
// Calculate coords for short tick length
x0 = sx * (100 + tl) + width*.75;
y0 = sy * (100 + tl) + height/2;
x1 = sx * 100 + width*.75;
y1 = sy * 100 + height/2;
// Draw tick
```

```
  tft.drawLine(x0, y0, x1, y1, BLACK);

  // Draw needle
  if (i==40){
    tft.drawLine(lnx0, lny0, lnx1, lny1, BLACK);
    float nsx = cos((lval/5 - 140) * 0.0174532925);
    float nsy = sin((lval/5 - 140) * 0.0174532925);
    uint16_t nx0 = nsx * (100 + tl) + width/4;
    uint16_t ny0 = nsy * (100 + tl) + height/2;
    uint16_t nx1 = nsx * 100 + width/4;
    uint16_t ny1 = nsy * 100 + height/2;
    tl=20;  // line length
    lnx0 = nsx * (1 + tl) + width/4;
    lny0 = nsy * (1 + tl) + height/2;
    lnx1 = nsx * 100 + width/4;
    lny1 = nsy * 100 + height/2;
    tft.drawLine(lnx0, lny0, lnx1, lny1, WHITE);

  // Right Needle
    tft.drawLine(rnx0+w2, rny0, rnx1+w2, rny1, BLACK);
    nsx = cos((rval/5 - 140) * 0.0174532925);
    nsy = sin((rval/5 - 140) * 0.0174532925);
    nx0 = nsx * (100 + tl) + width/4;
    ny0 = nsy * (100 + tl) + height/2;
    nx1 = nsx * 100 + width/4;
    ny1 = nsy * 100 + height/2;
    rnx0 = nsx * (1 + tl) + width/4;
    rny0 = nsy * (1 + tl) + height/2;
    rnx1 = nsx * 100 + width/4;
    rny1 = nsy * 100 + height/2;
    tft.drawLine(rnx0+w2, rny0, rnx1+w2, rny1, WHITE);
  }
}
  // Optional Text display of results
  tft.fillRect(0, height-40, width, 40, WHITE);
  tft.setTextColor(RED);
  tft.setCursor(10, height-30);
  tft.println(lval);
  tft.setCursor(width/2, height-30);
  tft.println(rval);
  delay(300);
}
```

For the Adafruit compatible LCD's change the first few lines of code to this:

```
// ADAFRUIT DRIVERS
#include <Adafruit_GFX.h>    // Core graphics library
#include <Adafruit_TFTLCD.h> // Hardware-specific library
#define LCD_CS A3 // Chip Select goes to Analog 3
#define LCD_CD A2 // Command/Data goes to Analog 2
#define LCD_WR A1 // LCD Write goes to Analog 1
#define LCD_RD A0 // LCD Read goes to Analog 0
#define LCD_RESET A4 // Can connect to Arduino's reset pin

Adafruit_TFTLCD tft(LCD_CS, LCD_CD, LCD_WR, LCD_RD,
LCD_RESET);

int width=480;  // scales to 320
int height=320;
uint16_t g_identifier;

int lnx0, lny0, lnx1, lny1;
int rnx0, rny0, rnx1, rny1;
```

MSGEQ7 Bar Graph

For this next project we will add the MSGEQ7 to the LCD screen. The LCD uses a lot of pins and we have to connect the MSGEQ7 to what pins are left over. As a result, the MSGEQ7 Strobe is on pin D2 and the Reset is on pin D3. The analog inputs are connected to pins A4 and A5. This is shown in the next schematic.

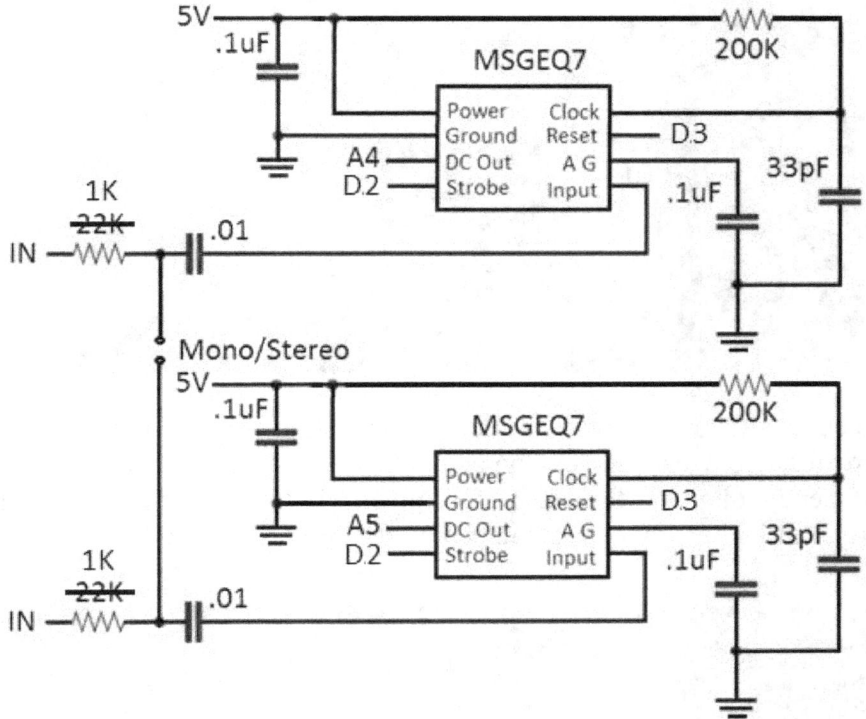

Since the LCD shield covers over all of the Arduino pins you will need to solder pins or wires on the back side of the Arduino to be able to connect the MSGEQ7.

Some versions of the LCD shield have a LM75 temperature sensor connected to A4 and A5, because they are the same as SCL and SDA. In that case you might want to remove or bend over the LCD shield pins for SCL and SDA, as they are the same as A4 and A5.

The pins you need to remove or bend are located three and four pins above D13 on the right side of the Arduino LCD Shield. The temperature sensor IC and the two pins to bend or remove, if the temperature sensor is there, are circled in the following picture.

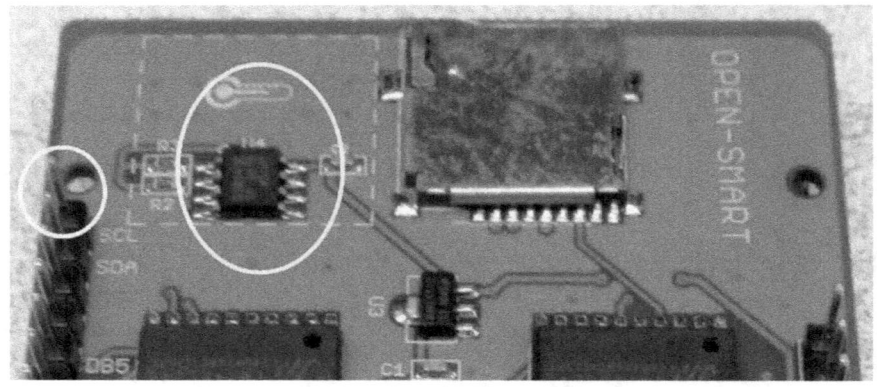

Since we only have seven frequency samples, and lots of available screen space, what I have done is to take two samples of each frequency, one after the other. This gives 14 samples per channel, for a total of 28 bars that are displayed on the screen.

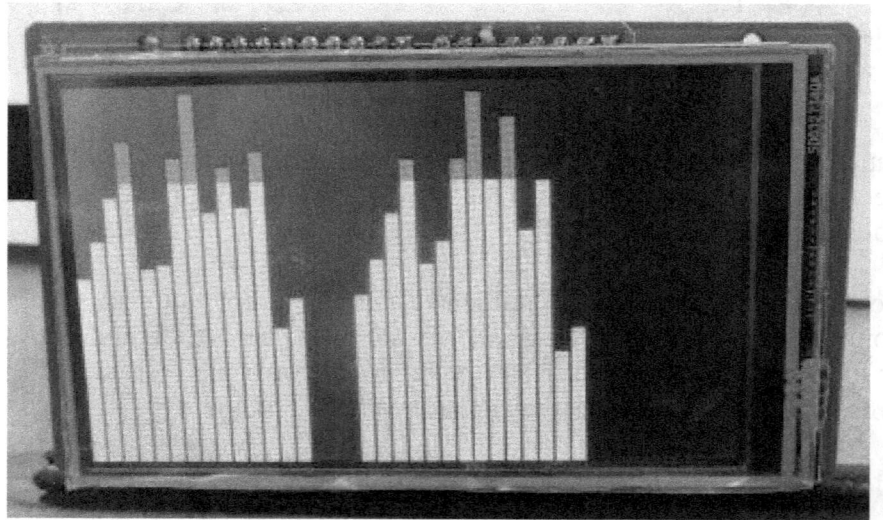

Here is the code for the MSGEQ7 with the LCD Screen. This version uses the MCUFRIEND_kbv tft driver to work with the 3.2 inch LCD screen.

```
// Open-Smart-MSGEQ7-2
// By Bob Davis
//
// Adafruit_TFTLCD LIBRARY MUST BE SPECIFICALLY
// CONFIGURED FOR EITHER THE TFT SHIELD
// SEE RELEVANT COMMENTS IN Adafruit_TFTLCD.h
```

```
// Some code by Open-Smart Team and Catalex Team
//
// Arduino IDE: 1.8.1
// Board: Arduino UNO R3 or Arduino Mega2560

// MSGEQ7 pins REMAPPED
#define PIN_STROBE 2
#define PIN_RESET 3
#define PIN_4 4 //analog input
#define PIN_5 5 //analog input
// band arrays
int left0[8];
int right0[8];
int left1[8];
int right1[8];

#include <MCUFRIEND_kbv.h>
MCUFRIEND_kbv tft;
int width=400;
int height=240;

//************************************************//
// The control pins for the LCD can be assigned to any pins
//
// TFT Breakout  -- Arduino UNO / Mega2560
// GND          -- GND
// 3V3          -- 3.3V
// CS           -- A3
// RS           -- A2
// WR           -- A1
// RD           -- A0
// RST          -- RESET
// LED          -- GND
// DB0-DB7      -- 8, 9, 10, 11, 4, 13, 6, 7

// Assign human-readable names to some common 16-bit color
values:
#define BLACK   0x0000
#define BLUE    0x001F
#define RED     0xF800
#define GREEN   0x07E0
#define CYAN    0x07FF
```

```
#define MAGENTA 0xF81F
#define YELLOW  0xFFE0
#define WHITE   0xFFFF

void readMSGEQ7() { //reset the chip
  digitalWrite(PIN_RESET, HIGH);
  digitalWrite(PIN_RESET, LOW);
  for(int band=0; band < 7; band++) {  //loop thru all 7 bands
    digitalWrite(PIN_STROBE,LOW);      // go to the next band
    delayMicroseconds(30);             // gather data
    left0[band] = analogRead(PIN_4)/4.3;  // store band reading
    right0[band] = analogRead(PIN_5)/4.3; // store band reading
    digitalWrite(PIN_STROBE,HIGH);     // reset the strobe pin
  }      // Second set of samples 200ms later
  for(int band=0; band < 7; band++) {  //loop thru all 7 bands
    digitalWrite(PIN_STROBE,LOW);      // go to the next band
    delayMicroseconds(30);             // gather data
    left1[band] = analogRead(PIN_4)/4.3;  // store band reading
    right1[band] = analogRead(PIN_5)/4.3; // store band reading
    digitalWrite(PIN_STROBE,HIGH);     // reset the strobe pin
  }
}

uint16_t g_identifier;

void setup(void) {
  pinMode(PIN_STROBE, OUTPUT);
  pinMode(PIN_RESET, OUTPUT);
  g_identifier = tft.readID(); //
  if (g_identifier == 0x00D3 || g_identifier == 0xD3D3) g_identifier
= 0x9481; // write-only shield
  if (g_identifier == 0xFFFF) g_identifier = 0x9341; // serial
  tft.begin(g_identifier);
  tft.setRotation(1);
  tft.setCursor(0, 0);
  tft.setTextColor(WHITE);
  tft.setTextSize(2);
  tft.fillRect(0, 0, width, height, BLACK);
}

void loop(void) {
  readMSGEQ7();
```

```
//   tft.println("DUAL MSGEQ7 ON A LCD SCREEN");
 int tred=height*3/4;
 int wstep=width/20;
 for (int b=0; b<8; b++){
  tft.fillRect(b*wstep, 0, 6, height, BLACK);
  tft.fillRect(b*wstep, height-left1[b], 6, height, GREEN);
  if (left1[b]>tred) {
    tft.fillRect(b*wstep, height-left1[b], 6, left1[b]-tred, RED);
  }
 }
  for (int b=0; b<8; b++){
   tft.fillRect(b*wstep+wstep/2, 0, 6, height, BLACK);
   tft.fillRect(b*wstep+wstep/2, height-left0[b], 6, height, GREEN);
   if (left0[b]>tred) {
    tft.fillRect(b*wstep+wstep/2, height-left0[b], 6, left0[b]-tred,
RED);
   }
 }
 for (int b=0; b<8; b++){
  tft.fillRect(b*wstep+width/2, 0, 6, height, BLACK);
  tft.fillRect(b*wstep+width/2, height-right1[b], 6, height, GREEN);
  if (right1[b]>tred) {
    tft.fillRect(b*wstep+width/2, height-right1[b], 6, right1[b]-tred,
RED);
   }
 }
  for (int b=0; b<8; b++){
   tft.fillRect(b*wstep+width/2+wstep/2, 0, 6, height, BLACK);
   tft.fillRect(b*wstep+width/2+wstep/2, height-right0[b], 6, height,
GREEN);
   if (right0[b]>tred) {
    tft.fillRect(b*wstep+width/2+wstep/2, height-right0[b], 6,
right0[b]-tred, RED);
   }
 }

 delay(100);
}
```

To work with the Adafruit compatible 2.4 inch and 3.5 inch screens
you will need to make some changes in the software and hardware.

The MSGEQ7 module will connect to D0, D1 and analog 4, analog 5 as in this next schematic.

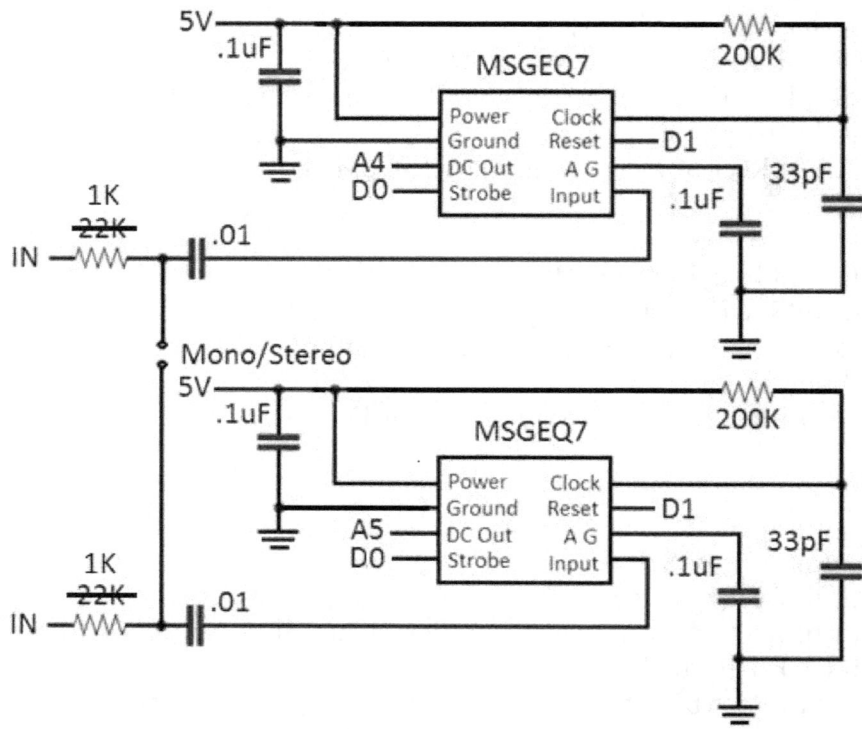

You need to bend over or remove the A4 (LCD_Reset) pin on the LCD so it can be used with the MSGEQ7 instead. You can leave A4/LCR Reset disconnected, that works with some LCD's but will cause problems with some screens. One solution is to tie the LCD Reset to the Arduino reset. The problem is that the Arduino reset signal is not available on the LCD screens connectors.

Another solution is to add a pull up resistor to the 3.3 or the 5-volt pin that is available on the LCD screens connectors. This solution is shown in the next picture. Any resistor form 100 ohms to 1000 ohms will do.

The first few lines of code need to be changed like this for the 2.4-inch and 3.5-inch screens:

```
// ADAFRUIT DRIVERS
#include <Adafruit_GFX.h>    // Core graphics library
#include <Adafruit_TFTLCD.h> // Hardware-specific library
#define LCD_CS A3 // Chip Select goes to Analog 3
#define LCD_CD A2 // Command/Data goes to Analog 2
#define LCD_WR A1 // LCD Write goes to Analog 1
#define LCD_RD A0 // LCD Read goes to Analog 0
// Fictional A6 to release A4 to be used by the MSGEQ7
#define LCD_RESET A6 // Can just connect to Arduino's reset pin
Adafruit_TFTLCD tft(LCD_CS, LCD_CD, LCD_WR, LCD_RD,
LCD_RESET);
uint16_t g_identifier = 0x9341;

// MSGEQ7 pins REMAPPED
#define PIN_STROBE 0
#define PIN_RESET 1
#define PIN_4 4 //analog input
#define PIN_5 5 //analog input
// band arrays
int left0[8];
int right0[8];
int left1[8];
int right1[8];
```

// Disable these lines and any line that changes g_identifier
//#include <MCUFRIEND_kbv.h>
//MCUFRIEND_kbv tft;

int width=480; // scales to 320 for 2.4"
int height=320; // use 240 for 2.4"

The Adafruit driver overrides the screen width and height settings. The screen size needs to be manually changed in the Adafruit driver library for the 3.5" screen. The "Adafruit_TFTLCD" file needs to be changed like the following example for the 3.5-inch screen, note that width and height are backwards:

// Manually override size for 3.5" screen
#define TFTWIDTH   320
#define TFTHEIGHT 480
// Default size settings for 2.4" screen
//#define TFTWIDTH   240
//#define TFTHEIGHT 320

Chapter 6

Arduino Addressable LED

Audio Projects

Addressable LED's are usually found in "Strips". They are also called names like "WS2812". The LED strip name can denote different LED IC versions, such as three wire and four wire devices.

Addressable LED's work by each LED reading its relevant info, 8 bits of red, 8 bits of green and 8 bits of blue, and then it sends the rest of the info on to the next addressable LED. After a short time of no data being sent to the LED's, all the LED's display the data that they have captured.

As you might guess, the timing of the data bits and the time there is no data, is critical to making the addressable LED strips work. When there is a time of no data, then the last data sent is displayed.

The LED strips come with adhesive on the back and I applied them to white corrugated plastic to make it solid. Then the data lines of the top two strips are connected to the Arduino. For the MSGEQ7 version all 14 LED strips are connected to D0 to D13. The power and ground lines are connected together and then connected to a five-volt 10 Amp power supply.

Up next is the schematic diagram showing how to wire up for two LED strips to the Arduino.

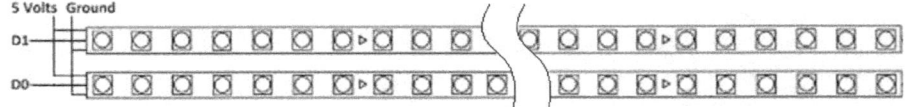

This picture is what a two-line addressable LED display should look like when running. The color changes from green to red at 3/4 of the maximum input level.

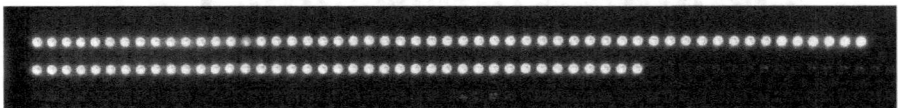

This is the code for two LED strips to take a reading from analog inputs A4 and A5 and display the reading on the LED strips. A4 and A5 are the analog inputs that are used for this sketch.

```
// Bob Davis Version for sending data to 2 Parallel WS2812 strings
// Removed Assembler and simplified the code
// Changed to 2 channel VU meter

// PORTD is Digital Pins 0-7 on the Uno change for other boards.
#define PIXEL_PORT  PORTD  // Port of the pin the pixels are
connected to
#define PIXEL_DDR   DDRD   // Port of the pin the pixels are
connected to
#define PIN_LEFT 4 //analog input
#define PIN_RIGHT 5 //analog input
// Create Variables
int maxled=64; // Maximum number of LED's per string
int left; int right;
int col;
int red; int green; int blue;

// Send the next set of 8 WS2812B encoded bits to the 8 pins.
// The delay timing is for an Arduino UNO.
void sendBitX8( uint8_t bits ) {
  PORTD= 0xFF;  // turn on
  PORTD= 0xFF;  // delay
  PORTD= 0xFF;  // delay
  PORTD= 0xFF;  // delay
  PORTD= 0xFF;  // delay (Add more for faster processors)
  PORTD= bits;  // send data
```

```
    PORTD= bits;  // delay
    PORTD= bits;  // delay
    PORTD= bits;  // delay
    PORTD= bits;  // delay
    PORTD= 0x00;  // Turn off;
}

void sendPixelRow( uint8_t row ) {
  // Send the bit 8 times down every row, 8 bits each for R,G,B
    for (int bit=0; bit<8; bit++){
      if (green==1)sendBitX8( row );
      else sendBitX8( 0x00 ); }
    for (int bit=0; bit<8; bit++){
      if (red==1)sendBitX8( row );
      else sendBitX8( 0x00 ); }
    for (int bit=0; bit<8; bit++){
      if (blue==1)sendBitX8( row );
      else sendBitX8( 0x00 ); }
  }

void setup() {
  PIXEL_DDR = 0xFF;    // Set all row pins to output
}
void loop() {
  red=0;
  green=1;
  blue=0;
  left=analogRead(4);
  right=analogRead(5);
    cli();                // No time for interruptions!
    for (int b=1; b<maxled; b++){  // number of bytes to send
    if (b >maxled*.75){ red=1; green=0;}
    else {red=0; green=1;}
    col=0;
    if (left-64 >= b) col=col+1;  // Send bytes as VU meter data
    if (right-64 >= b) col=col+2;  // Send bytes as VU meter data
    sendPixelRow(col);
    }
  sei();                // interrupts back on
  delay (50);
}
```

MSGEQ7 Bar Graph

This is the schematic for connecting 14 LED strips to an Arduino for use with a MSGEQ7 to display 14 Frequency VU levels.

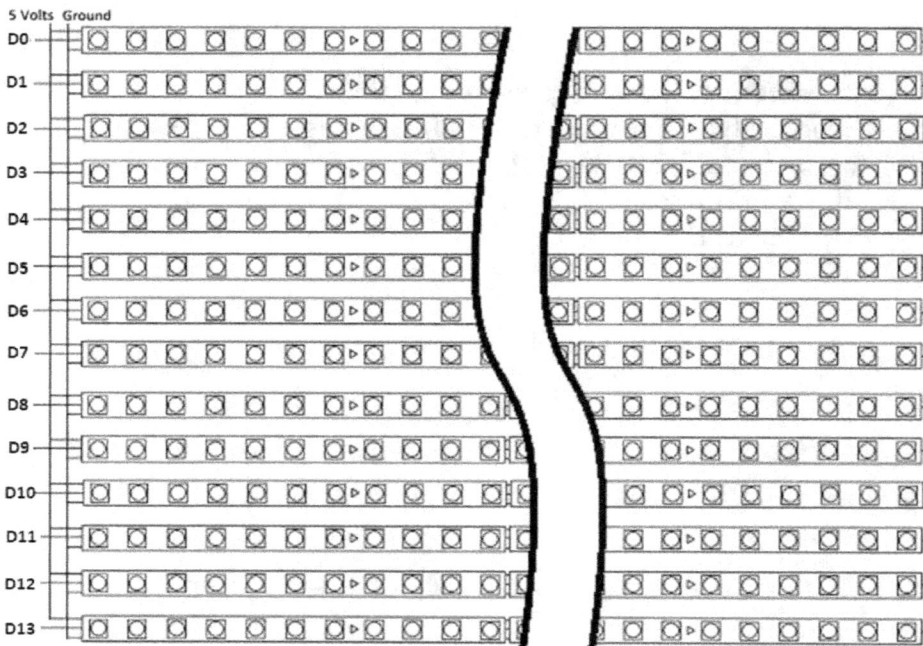

My software gives you up to 90, or more, LED's to the top of the sign board. Of course, you can go much bigger than that, if you want to, but the ceiling height can be a limiting factor. This picture is what the 14 LED strips looks like while displaying 14 frequency responses from the two MSGEQ7's.

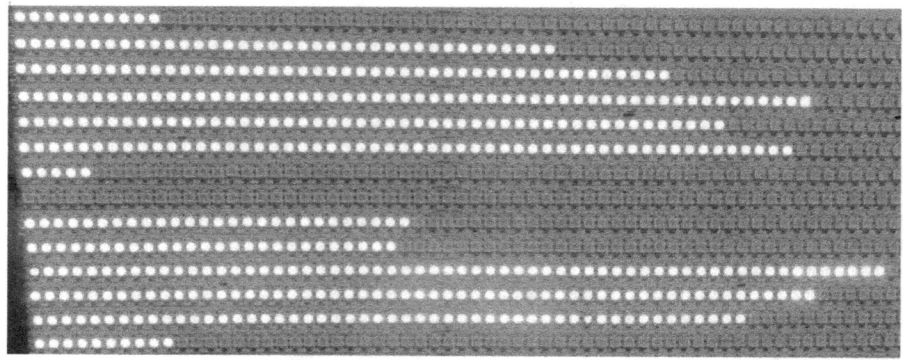

The wiring of this project is a bit of a mess. There are 14 power lines, 14 ground lines, 14 data lines and the six lines to the MSGEQ7

that all need to be connected up for it to work. This wiring mess can be seen in the next picture.

The MSGEQ7 is connected to A0 (strobe), A1 (reset), A4, and A5 (analog inputs). This is shown in the next schematic.

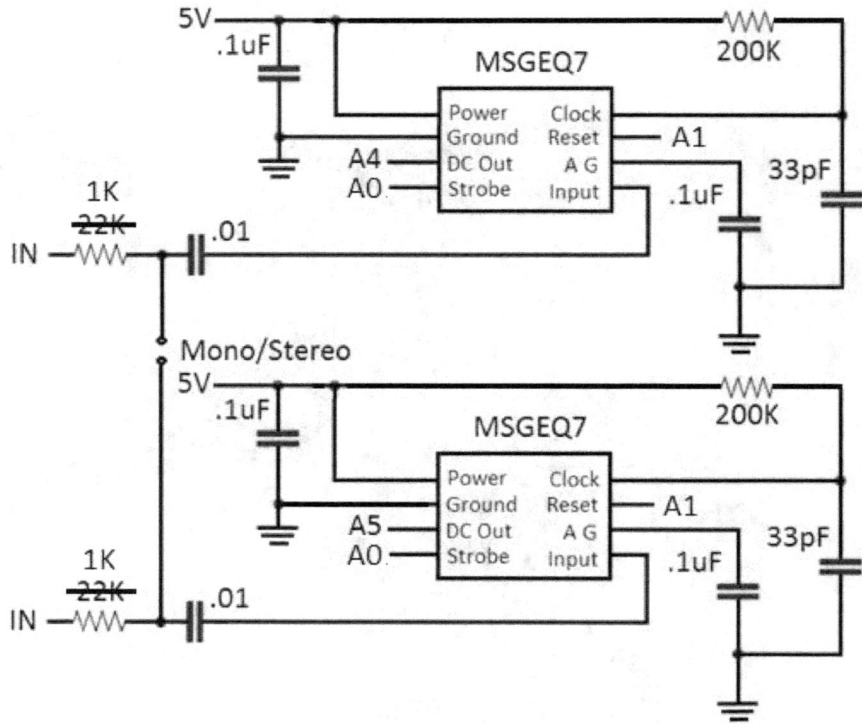

This is the code for the seven right and seven left channel MSGEQ7 version. You might need to adjust maxled to the amount of LED's you have in each LED strip.

What looks a little bit confusing is the code to turn on the binary bit to match with the column of the display. Instead of adding 1, 2, 4, 8 etc. I used hexadecimal notation: 0x0001, 0x0002, 0x0004, 0x0008, etc.

```
// BOB Davis Version for sending data to 14 Parallel WS2812 strings
// Removed Assembler and simplified the code
// Changed to VU meter with dual MSGEQ7's

// PORTD is Digital Pins 0-7 (Uses D0-D7=First 8 lines)
// POPRB is Digital Pins 8-13 (Uses D8-D13= Next 6 lines)
#define PIXEL_PORT   PORTD  // Port the pixels are connected to
#define PIXEL_DDR    DDRD   // Port the pixels are connected to
#define PIXEL_PORTB  PORTB  // Port the pixels are connected to
#define PIXEL_DDRB   DDRB   // Port the pixels are connected to

// MSGEQ7 pins uses 4 analog pins
```

```
#define PIN_STROBE A0
#define PIN_RESET A1
#define PIN_LEFT 4 //analog input
#define PIN_RIGHT 5 //analog input

int textb=128;  // brightness of text
int maxled=64; // Maximum number of LED's
//band arrays
int left[7]; int right[7];
int col=0;
// Set default colors
int red=textb; int green=textb; int blue=textb;

void readMSGEQ7() { //reset the chip
  digitalWrite(PIN_RESET, HIGH);
  digitalWrite(PIN_RESET, LOW);
  for(int band=0; band < 7; band++) { //loop thru all 7 bands
    digitalWrite(PIN_STROBE,LOW);      // go to the next band
    delayMicroseconds(30);             // gather data
    left[band] = analogRead(PIN_LEFT); // store left band reading
    right[band] = analogRead(PIN_RIGHT); // store right band reading
    digitalWrite(PIN_STROBE,HIGH);     // reset the strobe pin
  }
}

// Send the WS2812B encoded bits to the 14 pins.
// The delay timing is for an Arduino UNO.
void sendBitX8( byte bits, byte bith ) {
  PORTD= 0xFF;  // turn on
  PORTB= 0xFF;  // turn on
  PORTD= 0xFF;  // delay
  PORTD= 0xFF;  // delay
  PORTD= 0xFF;  // delay (add more for faster processors)
  PORTD= bits; // send data
  PORTB= bith; // send data
  PORTD= bits; // delay
  PORTD= bits; // delay
  PORTD= bits; // delay
  PORTD= 0x00; // Turn off;
  PORTB= 0x00; // Turn off;
}
```

```
void sendPixelRow( word row ) {
  // separate out upper and lower bytes from word
  byte rowh = row >> 8;
  // Send the bit 8 times down every row, each pixel is 8 bits/color
  int mask = 0x01;  // shifting mask to determine bit status
    for (int bit=8; bit>0; bit--){
      if (green & (mask << bit)) sendBitX8( row, rowh ); else
sendBitX8( 0x0000, 0x0000 ); }
    for (int bit=8; bit>0; bit--){
      if (red & (mask << bit)) sendBitX8( row, rowh ); else sendBitX8(
0x0000, 0x0000 ); }
    for (int bit=8; bit>0; bit--){
      if (blue & (mask << bit)) sendBitX8( row, rowh ); else
sendBitX8( 0x0000, 0x0000 ); }
 }

void setup() {
  PIXEL_DDR = 0xFF;    // Set all row pins to output
  PIXEL_DDRB = 0xFF;     // Set pins to output
  pinMode(PIN_RESET, OUTPUT); // reset
  pinMode(PIN_STROBE, OUTPUT); // strobe
}
void loop() {
  red=0; green=textb; blue=0;
  readMSGEQ7();                // collect samples
  cli();                 // No time for interruptions!
    for (int b=1; b<maxled; b++){ //maxled is number of bytes to send
      if (b >maxled*.75) {red=textb; green=0;}
      else {red=0; green=textb;}
      col=0;
      if (left[0]-64 >= b) col=col+0x0001;  // Set bits
      if (left[1]-64 >= b) col=col+0x0002;  // Set bits
      if (left[2]-64 >= b) col=col+0x0004;  // Set bits
      if (left[3]-64 >= b) col=col+0x0008;  // Set bits
      if (left[4]-64 >= b) col=col+0x0010;  // Set bits
      if (left[5]-64 >= b) col=col+0x0020;  // Set bits
      if (left[6]-64 >= b) col=col+0x0040;  // Set bits
      if (right[0]-64 >= b) col=col+0x0080;  // Set bits
      if (right[1]-64 >= b) col=col+0x0100;  // Set bits
      if (right[2]-64 >= b) col=col+0x0200;  // Set bits
      if (right[3]-64 >= b) col=col+0x0400;  // Set bits
      if (right[4]-64 >= b) col=col+0x0800;  // Set bits
```

```
    if (right[5]-64 >= b) col=col+0x1000;  // Set bits
    if (right[6]-64 >= b) col=col+0x2000;  // Set bits
    sendPixelRow(col);
     }
  sei();                    // interrupts back on
  delay (50);
}
```

8x8 Addressable LED Array

This 8x8 addressable LED array is marked as a CJMCU WS2812 Addressable LED Array. We will interface it with an Arduino Uno. The first addressable LED that I bought was an 8x8 LED array. I connected it up to an Arduino and loaded Adafruit's NeoPixel software into the Arduino and got NOTHING! Then, someplace, I read that the "Data in" and "Data out" labels are swapped. So, to get it to work, connect the displays "D out" pin to the Arduino D6 pin. It will then come to LIFE!

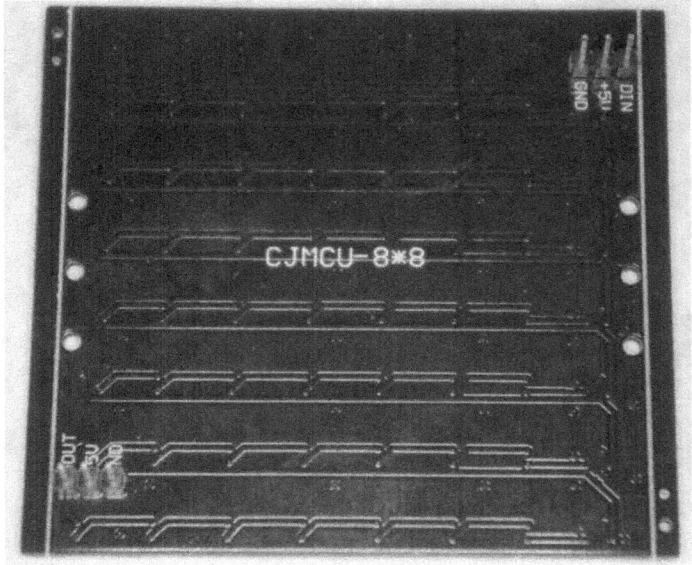

If you do not provide a separate power supply for the LED array after about a few minutes smoke will likely rise out of your Arduino. The voltage regulator will go up in smoke. To prevent the smoke, find a five-volt two amp AC adapter and use that to power the LED array.

Here is a picture of the 8x8 addressable LED array taken while displaying the status of the six analog inputs.

This next picture is the back side of the addressable LED array showing how to connect the array to the Arduino and to power. In this example, the AC adapter is also powering the Arduino. At the bottom of the picture, five volts power is coming in on a two-pin connector. A four-pin jumper is connected to the Arduino's five volts, D6, and ground. The fourth jumper pin is not used.

The next diagram shows how the 8x8 array is internally wired. Note that for this view the label will be upside down as in the drawing and the "in" jack is labeled "out".

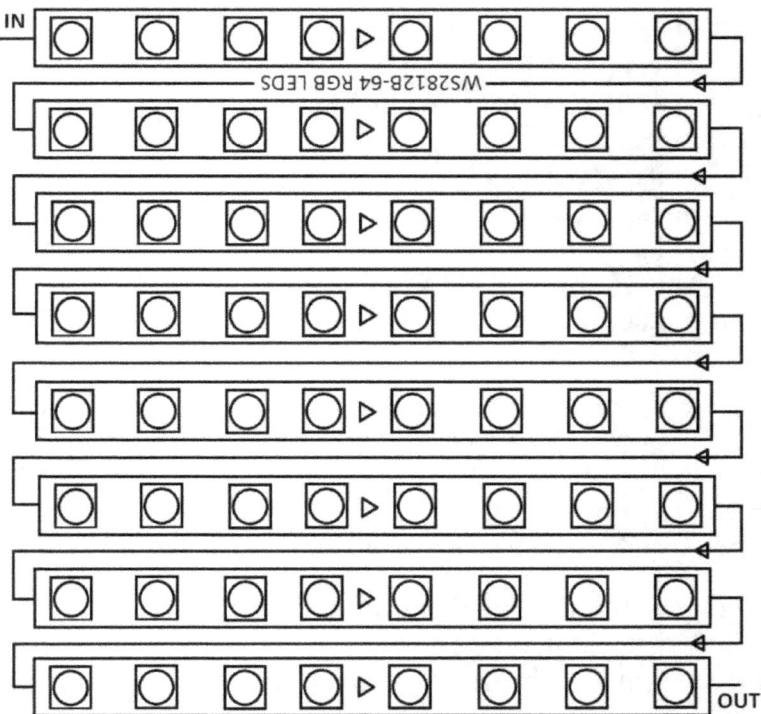

This next program can run one strip arranged into an array. It is set to use D6 but you can easily change that. The brightness and color are specified as three numbers separated by commas such as (4, 4, 4). The program will present six green bar graphs corresponding to the six analog inputs. The top of each bar graph will be red. This program could easily be modified to support larger displays by changing the maximum "led" value of LED's per row.

```
// Program for sending 6 Analogs to one WS2812 string
// By Robert Davis II
// Uses Digital Pin 6 change for other pins.
int LEDpin=6;
int MaxLed=64;
int DotBright=64; // sets your brightness
int led=8;  // Number of LED's per row

// The delay timing is for an Arduino UNO.
// Sorry digitalWrite is too slow to use.
void sendBit( uint8_t bits ) {
  PORTD= 0xF0;  // turn on
  PORTD= 0xF0;  // delay
  PORTD= 0xF0;  // delay
  PORTD= 0xF0;  // delay
  PORTD= 0xF0;  // delay (Add more for faster processors)
  PORTD= bits;  // send data
  PORTD= bits;  // delay
  PORTD= bits;  // delay
  PORTD= bits;  // delay
  PORTD= bits;  // delay
  PORTD= 0x00;  // Turn off;
}

void sendColor( int red, int green, int blue ) {
 // Send the bit 8 times down the row, each pixel is 8 bits each for
R,G,B
  int mask = 0x01; // shifting mask to determine bit status
    for (int bit=8; bit>0; bit--){
     if (green & (mask<<bit))sendBit( 0xF0 ); else sendBit( 0x00 ); }
    for (int bit=8; bit>0; bit--){
     if (red & (mask<<bit))sendBit( 0xF0 ); else sendBit( 0x00 ); }
    for (int bit=8; bit>0; bit--){
     if (blue & (mask<<bit))sendBit( 0xF0 ); else sendBit( 0x00 ); }
```

```
  }

void setup() {
  pinMode (LEDpin, OUTPUT);
  }

void loop() {
  // Read Analog Inputs
  int analog0 = analogRead(A0)/20.5;
  int analog1 = analogRead(A1)/20.5;
  int analog2 = analogRead(A2)/20.5;
  int analog3 = analogRead(A3)/20.5;
  int analog4 = analogRead(A4)/20.5;
  int analog5 = analogRead(A5)/20.5;
  cli();                    // No time for interruptions!
  for (int row=0; row<8; row++){  // Number of rows
    for (int l=0; l<led; l++){      // Number of LED's per row
      if (row==0){
        sendColor(0,0,0);
      }
      if (row==1){
        if (l<analog0) sendColor(0, DotBright, 0); // green
        if (l==analog0) sendColor(DotBright, 0, 0); // green
        if (l>analog0) sendColor(0,0,0);
      }
      if (row==2){
        if (l<analog1) sendColor(0, DotBright, 0); // green
        if (l==analog1) sendColor(DotBright, 0, 0); // green
        if (l>analog1) sendColor(0,0,0);
      }
      if (row==3){
        if (l<analog2) sendColor(0, DotBright, 0); // green
        if (l==analog2) sendColor(DotBright, 0, 0); // green
        if (l>analog2) sendColor(0,0,0);
      }
      if (row==4){
        if (l<analog3) sendColor(0, DotBright, 0); // green
        if (l==analog3) sendColor(DotBright, 0, 0); // green
        if (l>analog3) sendColor(0,0,0);
      }
      if (row==5){
        if (l<analog4) sendColor(0, DotBright, 0); // green
```

```
        if (l==analog4) sendColor(DotBright, 0, 0); // green
        if (l>analog4) sendColor(0,0,0);
      }
    if (row==6){
      if (l<analog5) sendColor(0, DotBright, 0); // green
      if (l==analog5) sendColor(DotBright, 0, 0); // green
      if (l>analog5) sendColor(0,0,0);
      }
    if (row==7){
      sendColor(0,0,0);
      }
    }
  //   delay(20);
  }
  sei();
  delay(1000);  // Turn on display
}
```

To add a MSGEQ7 follow the schematic from back when it was introduced. The Strobe is connected to D9 and the Reset is connected to D10. Analog inputs are on A4 and A5. Basically you just replace the analog0 with right[0] after adding the MSGEQ7 pin assignments and sub routine.

```
// Program for sending MSGEQ7 to one 8x8 WS2812 string
// By Robert Davis II
// Uses Digital Pin 6 change for other pins.
int LEDpin=6;
int MaxLed=64;
int DotBright=64; // sets your brightness
int led=8;  // Number of LED's per row
// MSGEQ7 stuff
int left[7];
int right[7];
#define PIN_STROBE 9
#define PIN_RESET 10
#define APIN0 4 //analog input
#define APIN1 5 //analog input
void readMSGEQ7() {
  //reset the chip
  digitalWrite(PIN_RESET, HIGH);
  digitalWrite(PIN_RESET, LOW);
```

```
  // Loop thru all 7 bands
  for(int band=0; band < 7; band++) {
    // Go to the next band
    digitalWrite(PIN_STROBE,LOW);
    // Delay for data to stabilize
    delayMicroseconds(30);
    // Store left band reading
    left[band] = analogRead(APIN0)/20;
    // Store right band reading
    right[band] = analogRead(APIN1)/20;
    // Reset the strobe pin
    digitalWrite(PIN_STROBE,HIGH);
    }
}

// The delay timing is for an Arduino UNO.
// Sorry digitalWrite is too slow to use.
void sendBit( uint8_t bits ) {
  PORTD= 0xF0;  // turn on
  PORTD= 0xF0;  // delay
  PORTD= 0xF0;  // delay
  PORTD= 0xF0;  // delay
  PORTD= 0xF0;  // delay (Add more for faster processors)
  PORTD= bits;  // send data
  PORTD= bits;  // delay
  PORTD= bits;  // delay
  PORTD= bits;  // delay
  PORTD= bits;  // delay
  PORTD= 0x00;  // Turn off;
}

void sendColor( int red, int green, int blue ) {
 // Send the bit 8 times down the row, each pixel is 8 bits each for
R,G,B
  int mask = 0x01;  // shifting mask to determine bit status
    for (int bit=8; bit>0; bit--){
     if (green & (mask<<bit))sendBit( 0xF0 ); else sendBit( 0x00 ); }
    for (int bit=8; bit>0; bit--){
      if (red & (mask<<bit))sendBit( 0xF0 ); else sendBit( 0x00 ); }
    for (int bit=8; bit>0; bit--){
      if (blue & (mask<<bit))sendBit( 0xF0 ); else sendBit( 0x00 ); }
  }
```

```
void setup() {
 pinMode(PIN_STROBE, OUTPUT);
 pinMode(PIN_RESET, OUTPUT);
 pinMode (LEDpin, OUTPUT);
 }

void loop() {
 // Read Analog Inputs
 readMSGEQ7();
 cli();                    // No time for interruptions!
 for (int row=0; row<8; row++){  // Number of rows
   for (int l=0; l<led; l++){     // Number of LED's per row
     if (row==0){
       sendColor(0,0,0);
     }
     if (row==1){
       if (l<right[0]) sendColor(0, DotBright, 0); // green
       if (l==right[0]) sendColor(DotBright, 0, 0); // green
       if (l>right[0]) sendColor(0,0,0);
     }
     if (row==2){
       if (l<right[1]) sendColor(0, DotBright, 0); // green
       if (l==right[1]) sendColor(DotBright, 0, 0); // green
       if (l>right[1]) sendColor(0,0,0);
     }
     if (row==3){
       if (l<right[2]) sendColor(0, DotBright, 0); // green
       if (l==right[2]) sendColor(DotBright, 0, 0); // green
       if (l>right[2]) sendColor(0,0,0);
     }
     if (row==4){
       if (l<right[3]) sendColor(0, DotBright, 0); // green
       if (l==right[3]) sendColor(DotBright, 0, 0); // green
       if (l>right[3]) sendColor(0,0,0);
     }
     if (row==5){
       if (l<right[4]) sendColor(0, DotBright, 0); // green
       if (l==right[4]) sendColor(DotBright, 0, 0); // green
       if (l>right[4]) sendColor(0,0,0);
     }
     if (row==6){
```

```
        if (l<right[5]) sendColor(0, DotBright, 0); // green
        if (l==right[5]) sendColor(DotBright, 0, 0); // green
        if (l>right[5]) sendColor(0,0,0);
      }
    if (row==7){
      if (l<right[6]) sendColor(0, DotBright, 0); // green
      if (l==right[6]) sendColor(DotBright, 0, 0); // green
      if (l>right[6]) sendColor(0,0,0);
      }
    }
  }
 sei();
 delay(200);  // Turn on display
}
```

Chapter 7

Arduino LED Panel

Audio Projects

LED panels come in many shapes and sizes. They are used to make modular LED signs. Modular signs are easy to repair, expand, and generally easier to work with. We will be using the popular 64 by 32 LED panel size for our projects.

We will be using the Adafruit RGB Matrix Shield for running the LCD Panels. This next picture shows optional modifications to connect E (Located between B and G2) to A4, using the lower yellow jumper wire. The upper yellow jumper wire is for the clock to go to D8 or to change it to D11 for use with the Arduino Mega.

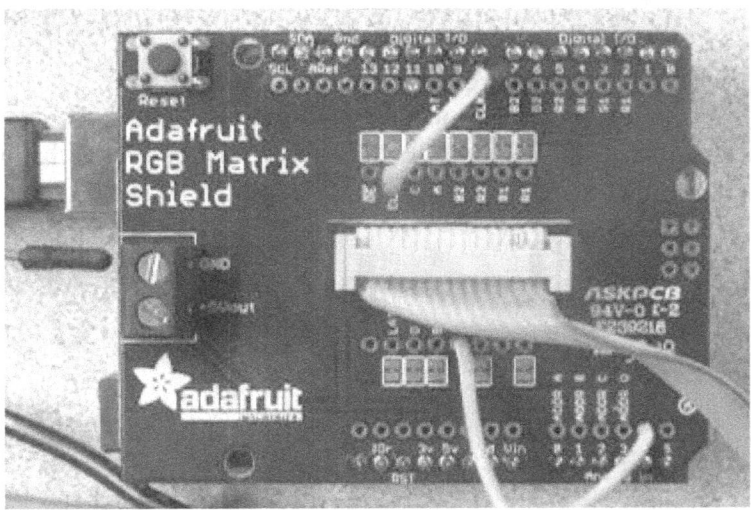

LED Panels follow a typical electronic design. There are shift registers that convert serial data to parallel data. There are two sets

of shift registers one for the top half and one for the lower half of the panel. In a three-color system there are three shift registers, one for each color, red, green and blue. The Data1 is called R1, G1, and B1 for the three-color panels. Then a row selection circuit selects what row is being displayed. The current row selected is usually identical between the top and bottom halves of the sign.

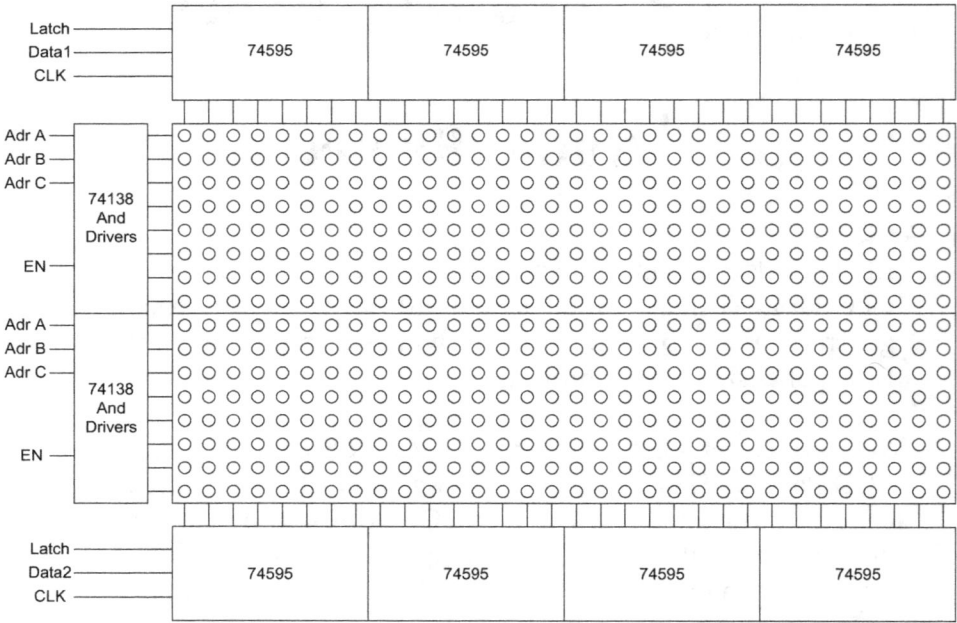

The LED panel will also require a five volt 10 or 12 amp power supply. Five-volt power connects to four pins located in the center of the back of the panel. A power cable and connector to match should come with the panel.

The Adafruit RGB Matrix Shield is inexpensive to buy but if you want to make your own here is the schematic diagram.

Arduino	Hub 75		Arduino
D2	R1 ○	○ G1	D3
D4	B1 ○	○ GND	GND
D5	R2 ○	○ G2	D6
D7	B2 ○	○ ADRE	A4
A0	ADRA ○	○ ADRB	A1
A2	ADRC ○	○ ADRD	A3
D8	CLK ○	○ LAT	D10
D9	OE ○	○ GND	GND

Our first LED panel project is to take the analog values on A4 and A5 and display their value and then display a bar graph showing their relative value as well as displaying the millivolt value in text form. This is shown in the next picture.

Here is the code for displaying two analog values. In the setup() we have to send some code to turn on the display and configure some control registers in the shift registers.

```
// 64x32 Uno LED panel Analog
// Fast Clock Mod
// 3/7/2019 by Bob Davis

// Port D assignments
// #define A   A0
// #define B   A1
// #define C   A2
// #define D   A3
// #define E   A4

// Port B assignments
// #define CLK 8
// #define OE  9
// #define LAT 10
//#define PIN_4 4 //analog input
//#define PIN_5 5 //analog input

int C12[16] = {0,1,1,1,1,1,1,1,1,1,1,1,1,1,1,1};
int C13[16] = {0,0,0,0,0,0,0,0,0,1,0,0,0,0,0,0};

#define MaxLed 64
byte BGC1=0x00;  // Background
byte FGC1=0x1C;  // Foreground
```

```
byte BGC2=0x00;  // Background
byte FGC2=0xE0;  // Foreground

#define PIXEL_PORT PORTD  // Port the pixels are connected to
#define PIXEL_DDR  DDRD   // D2-D7
#define ROW_PORT   PORTC  // Port the rows are connected to
#define ROW_DDR    DDRC   // A0-A5
#define CLK_PORT   PORTB  // Port the Clock/LE/OE are
connected to
#define CLK_DDR    DDRB   // D8-D10

char text1[10]=" ";
char text2[10]=" ";

// This font from http://sunge.awardspace.com/glcd-sd/node4.html
byte font[][7] = {
0x00,0x00,0x00,0x00,0x00,0x00,0x00, // ascii 32
0x00,0x00,0xfa,0x00,0x00,0x00,0x00, // !
0x00,0xe0,0x00,0xe0,0x00,0x00,0x00, // "
0x28,0xfe,0x28,0xfe,0x28,0x00,0x00, // #
0x00,0x34,0xfe,0x58,0x00,0x00,0x00, // $
0xc4,0xc8,0x10,0x26,0x46,0x00,0x00, // %
0x6c,0x92,0xaa,0x44,0x0a,0x00,0x00, // &
0x00,0xa0,0xc0,0x00,0x00,0x00,0x00, // '
0x00,0x38,0x44,0x82,0x00,0x00,0x00, // (
0x00,0x82,0x44,0x38,0x00,0x00,0x00, // )
0x10,0x54,0x38,0x54,0x10,0x00,0x00, // *
0x10,0x10,0x7c,0x10,0x10,0x00,0x00, // +
0x00,0x0a,0x0c,0x00,0x00,0x00,0x00, // ,
0x10,0x10,0x10,0x10,0x10,0x00,0x00, // -
0x00,0x06,0x06,0x00,0x00,0x00,0x00, // .
0x04,0x08,0x10,0x20,0x40,0x00,0x00, // /
0x7c,0x8a,0x92,0xa2,0x7c,0x00,0x00, // 0
0x00,0x42,0xfe,0x02,0x00,0x00,0x00, // 1
0x42,0x86,0x8a,0x92,0x62,0x00,0x00, // 2
0x84,0x82,0xa2,0xd2,0x8c,0x00,0x00, // 3
0x18,0x28,0x48,0xfe,0x08,0x00,0x00, // 4
0xe4,0xa2,0xa2,0xa2,0x9c,0x00,0x00, // 5
0x3c,0x52,0x92,0x92,0x0c,0x00,0x00, // 6
0x80,0x8e,0x90,0xa0,0xc0,0x00,0x00, // 7
0x6c,0x92,0x92,0x92,0x6c,0x00,0x00, // 8
0x60,0x92,0x92,0x94,0x78,0x00,0x00, // 9
```

```
0x00,0x6c,0x6c,0x00,0x00,0x00,0x00, // :
0x00,0x6a,0x6c,0x00,0x00,0x00,0x00, // ;
0x00,0x10,0x28,0x44,0x82,0x00,0x00, // <
0x28,0x28,0x28,0x28,0x28,0x00,0x00, // =
0x82,0x44,0x28,0x10,0x00,0x00,0x00, // >
0x40,0x80,0x8a,0x90,0x60,0x00,0x00, // ?
0x4c,0x92,0x9e,0x82,0x7c,0x00,0x00, // @
0x7e,0x90,0x90,0x90,0x7e,0x00,0x00, // A
0xfe,0x92,0x92,0x92,0x6c,0x00,0x00, // B
0x7c,0x82,0x82,0x82,0x44,0x00,0x00, // C
0xfe,0x82,0x82,0x82,0x7c,0x00,0x00, // D
0xfe,0x92,0x92,0x92,0x82,0x00,0x00, // E
0xfe,0x90,0x90,0x80,0x80,0x00,0x00, // F
0x7c,0x82,0x82,0x8a,0x4c,0x00,0x00, // G
0xfe,0x10,0x10,0x10,0xfe,0x00,0x00, // H
0x00,0x82,0xfe,0x82,0x00,0x00,0x00, // I
0x04,0x02,0x82,0xfc,0x80,0x00,0x00, // J
0xfe,0x10,0x28,0x44,0x82,0x00,0x00, // K
0xfe,0x02,0x02,0x02,0x02,0x00,0x00, // L
0xfe,0x40,0x20,0x40,0xfe,0x00,0x00, // M
0xfe,0x20,0x10,0x08,0xfe,0x00,0x00, // N
0x7c,0x82,0x82,0x82,0x7c,0x00,0x00, // O
0xfe,0x90,0x90,0x90,0x60,0x00,0x00, // P
0x7c,0x82,0x8a,0x84,0x7a,0x00,0x00, // Q
0xfe,0x90,0x98,0x94,0x62,0x00,0x00, // R
0x62,0x92,0x92,0x92,0x8c,0x00,0x00, // S
0x80,0x80,0xfe,0x80,0x80,0x00,0x00, // T
0xfc,0x02,0x02,0x02,0xfc,0x00,0x00, // U
0xf8,0x04,0x02,0x04,0xf8,0x00,0x00, // V
0xfe,0x04,0x18,0x04,0xfe,0x00,0x00, // W
0xc6,0x28,0x10,0x28,0xc6,0x00,0x00, // X
0xc0,0x20,0x1e,0x20,0xc0,0x00,0x00, // Y
0x86,0x8a,0x92,0xa2,0xc2,0x00,0x00,  // Z
};

void setup() {
  PIXEL_DDR = 0xFC;  // Set all pixel pins to output
  ROW_DDR = 0x0F;    // Set all row pins to output
  CLK_DDR = 0x0F;    // Set all CLK/LE/OE pins to output

  PORTB=0;
  // Send Data to control register 11
```

```
    for (int l=0; l<MaxLed; l++){
      int y=l%16;
      PORTD = 0x00;
      if (C12[y]==1) PORTD=0xFC;
        if (l>MaxLed-12){ PORTB=7; PORTB=6; }
        else{ PORTB=1; PORTB=0; }
      }

    PORTB=0;
    // Send Data to control register 12
    for (int l=0; l<MaxLed; l++){
      int y=l%16;
      PORTD = 0x00;
      if (C13[y]==1) PORTD=0xFC;
        if (l>MaxLed-13){ PORTB=7; PORTB=6; }
        else{ PORTB=1; PORTB=0; }
      }
    PORTB=0;
}

void loop() {
  for (int c=0; c<600; c++){
    int  va4 = analogRead(A4)/2; // Get the value
    int  va5 = analogRead(A5)/2; // Get the value
    sprintf(text1, "A4 = %03d", va4);
    sprintf(text2, "A5 = %03d", va5);

    // Select the Row
    for (int r=0; r<16; r++){
      for (int l=0; l<MaxLed; l++){
        int y=l%8; // remainder after division
        int pd1 = BGC1;
        int pd2 = BGC2;
        if (y < 6){
          if (r<8){
            if ((font[text1[l/8]-32][y] >> 8-r) & 0x01==1) pd1=FGC1;
            if ((font[text2[l/8]-32][y] >> 8-r) & 0x01==1) pd2=FGC2;
          }
          if (r==11 or r==12) {
            if (l<va4/8) {pd1=0x08;
              if (l >48) pd1=0x04;
            }
```

```
    if (l<va5/8) {pd2=0x40;
      if (l >48) pd2=0x20;
      }
    }
    }
    PORTD=pd1+pd2;
    }
    if (l>MaxLed-3){ PORTB=7; PORTB=6; }
    else{ PORTB=1; PORTB=0; }
    }
    PORTC=r;  // Update row
    PORTB=0;
  }
 }
}
```

MSGEQ7 Bar Graph

The MSGEQ7 LED panel version turns the bar graphs so that they
run the other way too on the LED panel. There is a version of this
display being sold on the Internet without the MSGEQ7. That
version creates the frequency bars in software only. The problem
with the software method is that it is not very accurate at all. There
are some web sites exposing this fakery.

I had to add pins to the RGB Matrix shield to be able to connect the
MSGEQ7's. You will need to add pins to the analog A4, A5, D9 and
D10, as well as to Power and ground as a minimum. I added pins to
all of the Arduino connections.

This next schematic shows how to connect the MSGEQ7's.

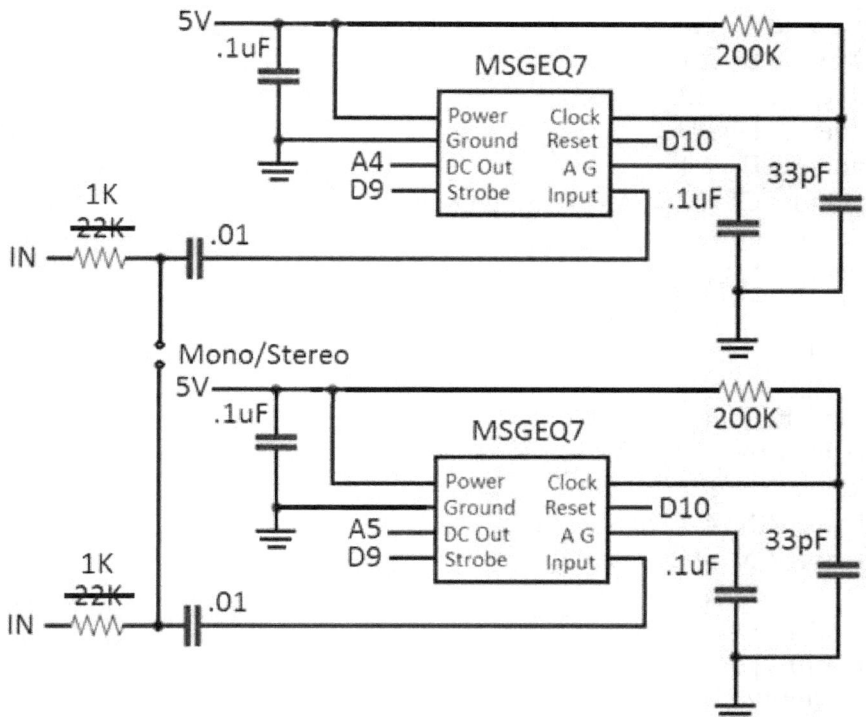

This is a picture of the LED panel running with the MSGEQ7

Here is the code for the MSGEQ7 bar graph. The LED color changes from green to red as you near the top of the LED array.

```
// 64x32 Uno LED panel with MSGEQ7
// Fast Clock Mod
// 3/7/2019 by Bob Davis
```

```
// #define A   A0  // Port D assignments
// #define B   A1
// #define C   A2
// #define D   A3
// #define E   A4
// #define CLK 8 // Port B assignments
// #define OE  9
// #define LAT 10
//#define PIN_4 4 //analog input
//#define PIN_5 5 //analog input

// MSGEQ7 pins REMAPPED
#define PIN_STROBE 11
#define PIN_RESET 12
#define PIN_4 4 //analog input
#define PIN_5 5 //analog input

//band arrays
int left0[7];
int right0[7];

int col=0;

void readMSGEQ7() { //reset the chip
  digitalWrite(PIN_RESET, HIGH);
  digitalWrite(PIN_RESET, LOW);
  for(int band=0; band < 7; band++) {  //loop thru all 7 bands
    digitalWrite(PIN_STROBE,LOW);      // go to the next band
    delayMicroseconds(30);             // gather data
    left0[band] = analogRead(PIN_4)-32;  // store band reading
    right0[band] = analogRead(PIN_5)-32;  // store band reading
    digitalWrite(PIN_STROBE,HIGH);     // reset the strobe pin
  }
}

int C12[16] = {0,1,1,1,1,1,1,1,1,1,1,1,1,1,1,1};
int C13[16] = {0,0,0,0,0,0,0,0,0,1,0,0,0,0,0,0};

#define MaxLed 64
byte BGC1=0x00; // Background
byte FGC1=0x1C; // Foreground
byte BGC2=0x00; // Background
```

```
byte FGC2=0xE0;  // Foreground

#define PIXEL_PORT PORTD  // Port the pixels are connected to
#define PIXEL_DDR  DDRD   // D2-D7
#define ROW_PORT   PORTC  // Port the rows are connected to
#define ROW_DDR    DDRC   // A0-A5
#define CLK_PORT   PORTB  // Port Clock/LE/OE connected to
#define CLK_DDR    DDRB   // D8-D10

void setup() {
  PIXEL_DDR = 0xFC;  // Set all pixel pins to output
  ROW_DDR = 0x0F;    // Set all row pins to output
  CLK_DDR = 0xFF;    // Set all CLK/LE/OE pins to output

  PORTB=0;
  // Send Data to control register 11
  for (int l=0; l<MaxLed; l++){
    int y=l%16;
    PORTD = 0x00;
    if (C12[y]==1) PORTD=0xFC;
      if (l>MaxLed-12){ PORTB=7; PORTB=6; }
      else{ PORTB=1; PORTB=0; }
    }
  PORTB=0;
  // Send Data to control register 12
  for (int l=0; l<MaxLed; l++){
    int y=l%16;
    PORTD = 0x00;
    if (C13[y]==1) PORTD=0xFC;
      if (l>MaxLed-13){ PORTB=7; PORTB=6; }
      else{ PORTB=1; PORTB=0; }
    }
  PORTB=0;
}

void loop() {
  readMSGEQ7();
  // Select the Row
  for (int r=0; r<16; r++){
    for (int l=0; l<MaxLed; l++){
      int pd1 = BGC1; int pd2 = BGC2;
      FGC1=0x08; FGC2=0x40;
```

```
if (l > MaxLed*.75){FGC1=0x04; FGC2=0x20; }
if (r==1 or r==2) {
  if (l<left0[0]/8) pd1=FGC1;
  if (l<right0[0]/8) pd2=FGC2;}
if (r==3 or r==4) {
  if (l<left0[1]/8) pd1=FGC1;
  if (l<right0[1]/8) pd2=FGC2;}
if (r==5 or r==6) {
  if (l<left0[2]/8) pd1=FGC1;
  if (l<right0[2]/8) pd2=FGC2;}
if (r==7 or r==8) {
  if (l<left0[3]/8) pd1=FGC1;
  if (l<right0[3]/8) pd2=FGC2;}
if (r==9 or r==10) {
  if (l<left0[4]/8) pd1=FGC1;
  if (l<right0[4]/8) pd2=FGC2;}
if (r==11 or r==12) {
  if (l<left0[5]/8) pd1=FGC1;
  if (l<right0[5]/8) pd2=FGC2;}
if (r==13 or r==14) {
  if (l<left0[6]/8) pd1=FGC1;
  if (l<right0[6]/8) pd2=FGC2;}
PORTD=pd1+pd2;
if (l>MaxLed-3){PORTB=7; PORTB=6;}
else{ PORTB=1; PORTB=0; }
}
PORTC=r;  // Update row
PORTB=0;  // Turn off latch, turn on display
}
}
```

Chapter 8

Arduino LED Sign

Conversions

You can use an old LED sign as a VU meter or spectrum analyzer.
I have interfaced an Arduino UNO to a Sunrise Systems LED sign as
well as a Silent Radio LED sign. The LED signs only need a 74138
and seven PNP driver transistors such as TIP127's. I am also using a
5-volt 5 amp regulated power supply to power the modified sign.

This is a picture of the Sunrise Systems controller that was removed.

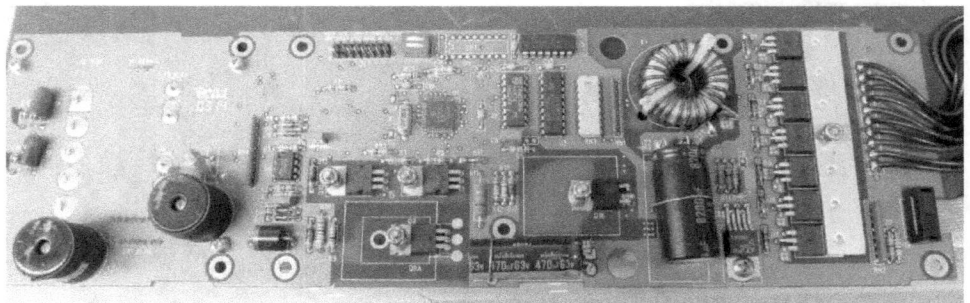

This is the silent Radio controller that is removed.

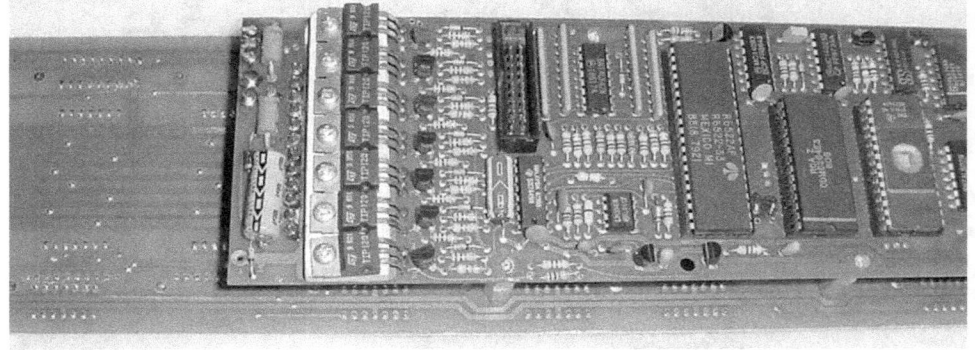

Most LED signs follow the same pattern. Basically, you have a lot of shift registers like the 74LS595 receiving "Data" and "Clock" signals and converting the serial data to parallel data. Then a "Latch" signal copies the data from the shift registers to output latches. To select the current row a 74LS138 or similar drives some power transistors to provide 5 volts power to the LED rows.

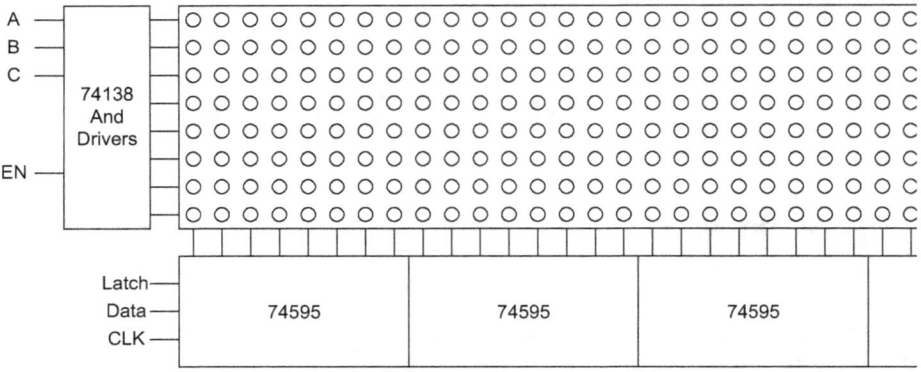

This next picture is a close up of the logic connection on the sunrise systems sign, the pins are five volts, Latch, Data, Clock, and Ground.

This next picture is a close up of the interface circuitry during an early test of the design.

This next picture is of a Silent Radio sign that was converted. When this conversion was done, I was not using the 74LS138 so it took less parts for the conversion.

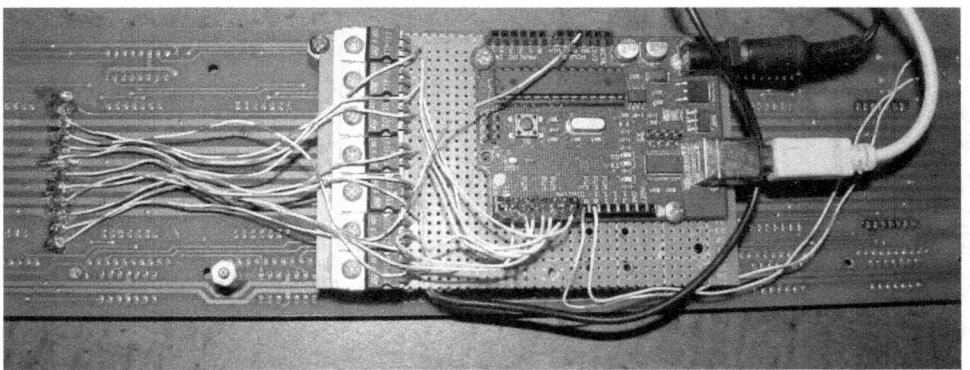

These next two picture shows some text being displayed on a converted LED sign. Instead of just seeing 96 LED columns, the software sees 16 characters and 6 columns per character.

This is the sign displaying an audio spectrum analysis with a MSGEQ7.

This is the schematic of the 74LS138 and PNP transistor LED sign interface. The 74138 outputs are low when selected so it has to be inverted back to high by the PNP driver transistors. The Arduino in the schematic is only the AT Mega IC chip but a full-size Arduino could easily be used as well.

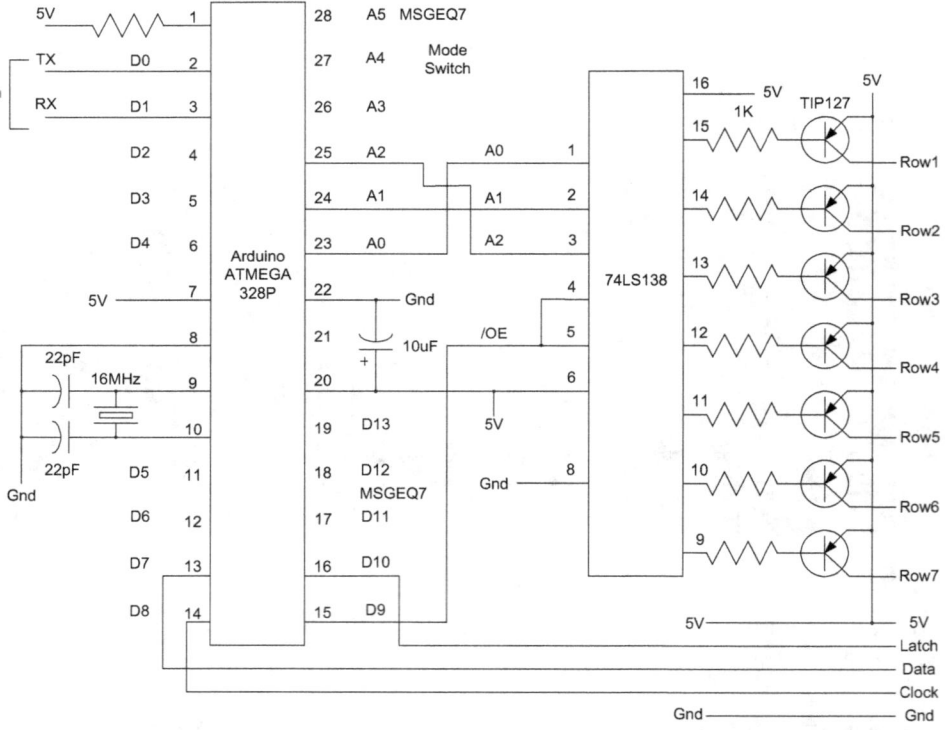

Adding the MSGEQ7

There is only one MSGEQ7 for this design because there are only 7 or 8 LED rows of LED's. Two signs can be used one on each side of the room for stero.

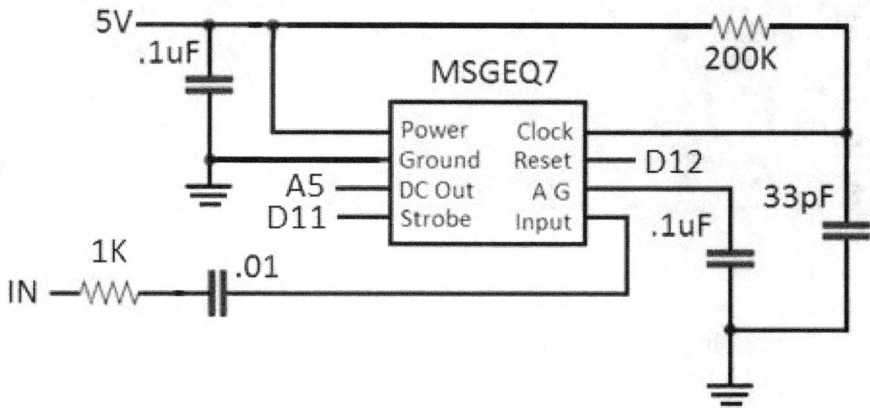

This is what the new sign control board looks like. It looks like it is rather empty. It used an AT Mega IC with the Arduino code in it so there is no Arduino circuit board needed. It uses a 5-volt 3-amp AC power adapter instead of having the voltage regulators located on the controller board.

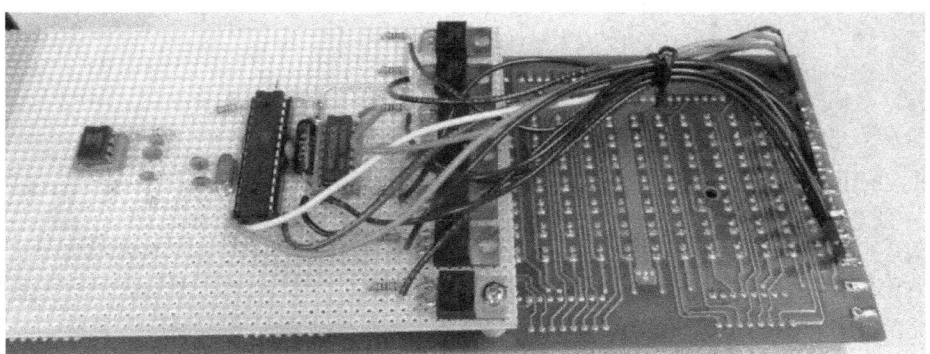

The converted sign can also have a Bluetooth interface so you can program it from your phone. This picture shows what is inside the new controller with Bluetooth.

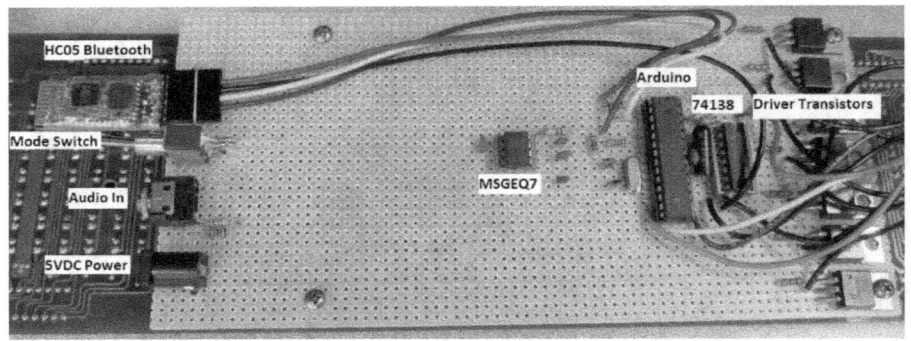

Here is the code to make it work with a 74LS138, seven 1K ohm resistors, seven TIP127's, and a MSGEQ7. You can optionally add the HC-05 Bluetooth interface to update the text with your phone. A mode switch on A4 selects between text and audio display.

```
// 7x96 Uno LED Array driver
// Does both text and MSGEQ7 Spectrum Analyzer
// Fast Clock Mod-Direct port writes
// 6/4/2019 by Bob Davis

// #define A   A0  74138 pin1
// #define B   A1  74138 pin2
// #define C   A2  74138 pin3
// #define CLK 8 // Port B assignments
// #define OE  9 // 74138 pins 4 and 5
// #define LAT 10// Latch

#define PIXEL_PORT PORTD  // Port the pixels are connected to
#define PIXEL_DDR  DDRD   // D2-D7
#define ROW_PORT   PORTC  // Port the rows 74138 are connected to
#define ROW_DDR    DDRC   // A0-A5
#define CLK_PORT   PORTB  // Port the Clock/OE/LE are connected to
#define CLK_DDR    DDRB   // D8-D13
#define MaxLed 96

// MSGEQ7 pins REMAPPED
#define PIN_STROBE 11
#define PIN_RESET 12
#define PIN_4 4 //analog input
#define PIN_5 5 //analog input
```

```
//band arrays
int left[7];
int right[7];
//int col=0;

String text="SIGN BY BOB DAVIS        ";

// This font from http://sunge.awardspace.com/glcd-sd/node4.html
byte font[][7] = {
0x00,0x00,0x00,0x00,0x00,0x00,0x00, // Sp
0x00,0x00,0xfa,0x00,0x00,0x00,0x00, // !
0x00,0xe0,0x00,0xe0,0x00,0x00,0x00, // "
0x28,0xfe,0x28,0xfe,0x28,0x00,0x00, // #
0x00,0x34,0xfe,0x58,0x00,0x00,0x00, // $
0xc4,0xc8,0x10,0x26,0x46,0x00,0x00, // %
0x6c,0x92,0xaa,0x44,0x0a,0x00,0x00, // &
0x00,0xa0,0xc0,0x00,0x00,0x00,0x00, // '
0x00,0x38,0x44,0x82,0x00,0x00,0x00, // (
0x00,0x82,0x44,0x38,0x00,0x00,0x00, // )
0x10,0x54,0x38,0x54,0x10,0x00,0x00, // *
0x10,0x10,0x7c,0x10,0x10,0x00,0x00, // +
0x00,0x0a,0x0c,0x00,0x00,0x00,0x00, // ,
0x10,0x10,0x10,0x10,0x10,0x00,0x00, // -
0x00,0x06,0x06,0x00,0x00,0x00,0x00, // .
0x04,0x08,0x10,0x20,0x40,0x00,0x00, // /
0x7c,0x8a,0x92,0xa2,0x7c,0x00,0x00, // 0
0x00,0x42,0xfe,0x02,0x00,0x00,0x00, // 1
0x42,0x86,0x8a,0x92,0x62,0x00,0x00, // 2
0x84,0x82,0xa2,0xd2,0x8c,0x00,0x00, // 3
0x18,0x28,0x48,0xfe,0x08,0x00,0x00, // 4
0xe4,0xa2,0xa2,0xa2,0x9c,0x00,0x00, // 5
0x3c,0x52,0x92,0x92,0x0c,0x00,0x00, // 6
0x80,0x8e,0x90,0xa0,0xc0,0x00,0x00, // 7
0x6c,0x92,0x92,0x92,0x6c,0x00,0x00, // 8
0x60,0x92,0x92,0x94,0x78,0x00,0x00, // 9
0x00,0x6c,0x6c,0x00,0x00,0x00,0x00, // :
0x00,0x6a,0x6c,0x00,0x00,0x00,0x00, // ;
0x00,0x10,0x28,0x44,0x82,0x00,0x00, // <
0x28,0x28,0x28,0x28,0x28,0x00,0x00, // =
0x82,0x44,0x28,0x10,0x00,0x00,0x00, // >
0x40,0x80,0x8a,0x90,0x60,0x00,0x00, // ?
0x4c,0x92,0x9e,0x82,0x7c,0x00,0x00, // @
```

```
0x7e,0x90,0x90,0x90,0x7e,0x00,0x00, // A
0xfe,0x92,0x92,0x92,0x6c,0x00,0x00, // B
0x7c,0x82,0x82,0x82,0x44,0x00,0x00, // C
0xfe,0x82,0x82,0x82,0x7c,0x00,0x00, // D
0xfe,0x92,0x92,0x92,0x82,0x00,0x00, // E
0xfe,0x90,0x90,0x80,0x80,0x00,0x00, // F
0x7c,0x82,0x82,0x8a,0x4c,0x00,0x00, // G
0xfe,0x10,0x10,0x10,0xfe,0x00,0x00, // H
0x00,0x82,0xfe,0x82,0x00,0x00,0x00, // I
0x04,0x02,0x82,0xfc,0x80,0x00,0x00, // J
0xfe,0x10,0x28,0x44,0x82,0x00,0x00, // K
0xfe,0x02,0x02,0x02,0x02,0x00,0x00, // L
0xfe,0x40,0x20,0x40,0xfe,0x00,0x00, // M
0xfe,0x20,0x10,0x08,0xfe,0x00,0x00, // N
0x7c,0x82,0x82,0x82,0x7c,0x00,0x00, // O
0xfe,0x90,0x90,0x90,0x60,0x00,0x00, // P
0x7c,0x82,0x8a,0x84,0x7a,0x00,0x00, // Q
0xfe,0x90,0x98,0x94,0x62,0x00,0x00, // R
0x62,0x92,0x92,0x92,0x8c,0x00,0x00, // S
0x80,0x80,0xfe,0x80,0x80,0x00,0x00, // T
0xfc,0x02,0x02,0x02,0xfc,0x00,0x00, // U
0xf8,0x04,0x02,0x04,0xf8,0x00,0x00, // V
0xfe,0x04,0x18,0x04,0xfe,0x00,0x00, // W
0xc6,0x28,0x10,0x28,0xc6,0x00,0x00, // X
0xc0,0x20,0x1e,0x20,0xc0,0x00,0x00, // Y
0x86,0x8a,0x92,0xa2,0xc2,0x00,0x00, // Z
};

void readMSGEQ7() { //reset the chip
  digitalWrite(PIN_RESET, HIGH);
  digitalWrite(PIN_RESET, LOW);
  for(int band=0; band < 7; band++) { //loop thru all 7 bands
   digitalWrite(PIN_STROBE,LOW);      // go to the next band
   delayMicroseconds(30);            // gather data
//   left[band] = analogRead(PIN_4)-24;  // store band reading
   right[band] = analogRead(PIN_5)-24; // store band reading
   digitalWrite(PIN_STROBE,HIGH);     // reset the strobe pin
  }
}

void setup() {
  Serial.begin(9600);
```

```
Serial.println("Arduino is ready");
PIXEL_DDR = 0xFC;  // Set all pixel pins to output
ROW_DDR = 0x0F;    // Set all row pins to output except analog
inputs
CLK_DDR = 0xFF;    // Set all CLK/LE/OE pins to output
}

void loop() {
  if (Serial.available() > 0) text=Serial.readString();
  // Pad length to 16 characters
  for(int i = text.length(); i < 16; i++){
    text += ' '; }
  text.toUpperCase();
//  Serial.print(text);
  if (analogRead(PIN_4) < 100) {
    // Select the Row
    for (int r=0; r<8; r++){
      // select the character
      for (int ch=0; ch<16; ch++){
        // select the column within character
        for (int c=0; c<6; c++){
          PORTD = 0x00;
          if ((font[text[ch]-32][c] >> r+1) & 0x01==1) PORTD=0xF0;
          PORTB=5; PORTB=4;  // Toggle clock
        }
      }
      // row is done so display it
      PORTC=r;  // Update row
      PORTB=0x00;
    }
  }
  if (analogRead(PIN_4) > 100) {  // Read switch
    readMSGEQ7();
    // Select the Row
    for (int r=0; r<7; r++){
      for (int l=0; l<MaxLed; l++){
        if (l<right[r]/8) PORTD=0xF0;
        else PORTD = 0x00;
        PORTB=5; PORTB=4;  // toggle clock
      }
      PORTC=r;  // Update row
      PORTB=0x00;  // Turn off latch, turn on display
```

```
            delay(1); // Slow down analog display
        }
      }
   }
```

Bibliography

Running 8 LED strips in parallel using machine language to control the timing: (I use repeated commands as time delays instead)
https://wp.josh.com/2016/05/20/huge-scrolling-arduino-led-sign/

The WS2812 Specification sheet:
http://www.world-semi.com/

The MSGEQ7 Specification sheet:
http://www.mix-sig.com/index.php/msgeq7-seven-band-graphic-equalizer-display-filter

The HC06 Bluetooth Module Specifications:
https://www.olimex.com/Products/Components/RF/BLUETOOTH-SERIAL-HC-06/resources/hc06.pdf

Note: It is best to use a 5-volt adapter board with the Bluetooth module.

www.ingramcontent.com/pod-product-compliance
Lightning Source LLC
Chambersburg PA
CBHW080850220526
45467CB00008B/2455

*9 7 8 1 7 0 4 3 3 0 9 0 7 *